MODERN
ELEMENTARY GEOMETRY

Prentice-Hall International, Inc., London
Prentice-Hall of Australia, Pty. Ltd., Sydney
Prentice-Hall of Canada, Ltd., Toronto
Prentice-Hall of India Private Limited, New Delhi
Prentice-Hall of Japan, Inc., Tokyo

MODERN
ELEMENTARY
GEOMETRY

James M. Moser

University of Wisconsin

Prentice-Hall, Inc., Englewood Cliffs, New Jersey

Current printing (last digit):   10 9 8 7 6 5 4 3 2 1

13-593186-X

Library of Congress Catalog Card Number 78-137660

Printed in the United States of America

PREFACE

Modern Elementary Geometry is a content matter text on the subject of elementary Euclidean geometry. It has been prepared along the guidelines set up by the CUPM in its Level I recommendations for the training of teachers. While the text is appropriate for both pre-service and in-service teachers, it is also suitable for a first course in geometry at the college level as well as for anyone who wishes to learn about geometry. While the use of the text presupposes a prior course or courses in the development of the real number system, it does not require any prior formal training in geometry. It does, however, offer material of interest to those who may have had a course in geometry in high school.

It is the intent of the author to give the reader some feeling for the way that mathematics is developed. This includes a large amount of the inductive, exploratory approach to the subject. To this end, the text material and the exercises are "open-ended" to some degree. Rather than be an endless exposition of facts, the book is designed to force the reader to make conjectures on his own and then try to "prove" them in an informal way. The user of the book, whether teacher or student, is advised that some of the exercises are deliberately constructed to be misleading or impossible to answer. The author has the wish to make the reader engage in some thought and to ask questions himself, rather than "turn the crank" to get a lot of answers.

The level of rigor is intentionally set low to begin with and is gradually increased as the reader progresses through the material. It is the opinion of the author that too early an introduction to rigid, formal logic dampens the enthusiasm and desire of the student to become involved in the subject matter. It is the intent to build up a fairly substantial body of factual matter pertaining to geometry and then to examine the logical structure of the subject when the student has sufficient "tools" with which to work.

The latter part of the text has a greater emphasis upon the deductive point of view. Deductive logic and proof has a legitimate place in mathematics and no college level student of the subject should be allowed to miss an exposure to this side of mathematics.

Whenever possible throughout the text, the attempt has been made to tie together geometric, arithmetic, and algebraic ideas. While the specific aim is to teach the student something about Euclidean geometry, the general purpose is to give him a better understanding of mathematics and how it may be applied to our everyday lives.

The material has been class tested in a preliminary form and the present text includes many revisions suggested by students and colleagues. In particular the author wishes to thank Jean Ferris, Jerry P. Becker, Jeremy Kilpatrick, George A. Spooner, Bruce Meserve, Burton W. Jones, Earl Hasz, and Richard Roth. But most of all, he extends sincerest thanks and genuine affection to the students of Mathematics 121, who suffered through the preliminary versions of this text.

JAMES M. MOSER

CONTENTS

7. PROOF   146

8. PARALLELS AND PARALLELOGRAMS   178

9. AREA OF PLANE FIGURES   199

10. GEOMETRY IN THREE DIMENSIONS   217

MODERN
ELEMENTARY GEOMETRY

1

INTRODUCTION

Geometry is a subject that has intrigued man for a long time. Since he found himself living in a world filled with physical objects having what we would call **geometrical properties** it is not surprising that he has attempted to learn about and understand these objects and their properties. Notions such as size, shape, form, position, location, round, square, distance were considered. Even today, children of all ages are fascinated by the wonderful geometric shapes of snowflakes, seashells, honeycombs, starfish, spiders' webs, leaves on a tree, to name but a few.

The word "geometry" itself comes from several Greek words that literally mean "earth measure" in much the same way that the word "geology" means "earth study."

1.1 GEOMETRY AS A MATHEMATICAL MODEL OF SPACE

Since our world is filled with objects that can be characterized by saying they have geometric properties, it is quite natural that we should study geometry so that we might better understand the world around us. Ge-

1

ometry is the study of spatial relationships.   The geometry that we study, then, should be the science that gives us a good source of information about our world.   In the view of many persons, the geometry that does this best is *Euclidean geometry*, which is a formal study of the subject based upon the works of Euclid, the famous mathematician who lived around 300 B.C. Euclid took the then existing knowledge of geometry and organized it into a formal system of mathematics, developed in what he thought was a logical and natural fashion.   His work, called the *Elements*, set forth this organized system of geometrical knowledge.   The *Elements* (or adaptations thereof) is perhaps one of the most widely read set of books ever written.

As a mathematical system, Euclidean geometry is based on mathematical abstractions.   This idea of an abstraction must be absolutely clear to the student of geometry before he proceeds any further.   An **abstraction** is an idea that exists only in the mind.   Perhaps the best example of a mathematical abstraction is the idea of number.   A young child learns very early what it means to play with sets of objects that have two elements.   There may be two pull-toys, two stuffed animals, two parents, two fingers, etc.   From these, he learns the concept or idea of "two" or "two-ness" as being the property that these sets have in common.   Yet, the concept really exists only in his mind.   The symbol "2" or "II" (if he happened to be a Roman child of long ago) is not the number at all but merely a physical mark used to represent or call to mind the idea of "two."   These representations are called **numerals**.   The familiar algorithms or arithmetic, such as

$$
\begin{array}{r}
241 \\
+\,196 \\
\hline
437
\end{array}
\qquad
\begin{array}{r}
253 \\
-\,127 \\
\hline
126
\end{array}
\qquad
\begin{array}{r}
23 \\
6\,)\,\overline{138} \\
12 \\
\hline
18 \\
18 \\
\hline
\end{array}
$$

are simply mathematical models written on paper that represent an abstract operation that can be performed on the numbers.   Of course, in certain mathematical systems one can operate entirely on the level of abstract ideas and not consider the system as a mathematical model of some real object.

In the same way, the cover of this book, the page on which you are now concentrating, the shape of the door or window, the shape of a picture on the wall may all remind you of a rectangle.   In Euclidean geometry, a rectangle is an abstraction that can be properly defined as a certain set of points with given properties.   As such, it serves as an excellent mathematical model of those things mentioned above that have in common the property that they are *rectangular* in shape. However, by studying the abstraction or mathematical model in a strictly intellectual way, certain facts and relationships that have direct applications to the real object may be learned. Not only is this of theoretical interest, but it is also of practical utility. This idea of a mathematical model works in other areas besides geometry and is a good example of how mathematics has been of use to mankind. The strength of the mathematical model approach depends on how faithful the definitions of the abstraction are to the real objects that they are to represent.

Mention should be made at this point of the use of diagrams and pictures in the study of geometry. Often, in an attempt to study more closely the relationships of a geometrical object, a picture of it is drawn. The picture is usually drawn with close attention paid to the manner in which the mathematical abstraction is defined. This picture is not an abstraction; it is a real object, having three dimensions, even though the third dimension of thickness might be microscopic in size. The picture need not be thought of as a real object, however, but merely as a bridge to enable a reader to think more clearly about both the abstraction and the real object that it is supposed to represent. Try to remember it in this light; also remember that a picture is subject to the vagaries and errors of the one who draws it as well as the one who beholds it.

1.2   GEOMETRY AS A MATHEMATICAL SYSTEM

Chances are that if a former student of geometry were asked what he learned during his study of the subject, he might reply: "Nothing." Being duly surprised, the interrogator might continue with a question to the effect that if he learned nothing, then what in the world did he do all year in the class? To which the student might very well retort: "Proofs."

What this hypothetical and hopefully untrue situation suggests is the fact that one of the purposes of the study of geometry is to teach the nature of proof and what is commonly known as a **mathematical system.** Basically, a mathematical system is one in which certain statements are proven to be true on the basis of valid deductive arguments that, in turn, are based on previously proven statements and established definitions. Thus, a certain logical chain is built up, preferably from a very small initial set of statements and definitions. It does not take too much reflection to realize that these beginnings of the system must come from some place other than other statements and definitions. This point of beginning is a set of undefined terms and a set of unproven statements called **postulates.** To some, the smaller the set of beginning concepts, the more elegant the system. A concern for elegance is somewhat irrelevant to the discussion however. What is important is an understanding of the nature of a mathematical system. A more thorough coverage of the nature of proof and the development of a mathematical system will be given in later chapters.

The subject matter of a mathematical system may be quite abstract and have little relationship to actual events or objects, but it is necessary that the system be logically sound and validly developed. However, those systems that are based on real objects and known physical relationships tend to have a greater pertinence to the student. Such is the case of Euclidean geometry. Its great strength lies in its close agreement with our ordinary perceived space experiences. This is not to demean its systematic and logical development, which is all the more amazing when one considers the time when it was originally developed (300 B.C.). The geometry of Euclid has come down through the ages relatively unchanged and has been for many persons their only exposure to a study of a mathematical system. And what a classical study it is! If for no other reason, Euclidean geometry should be studied for its historical and cultural value.

1.3 FORMAL AND INFORMAL GEOMETRY

Most of us know what a doctor does, and if we were asked to tell what an architect, or bricklayer, or bus driver did all day, we would not have too much difficulty in answering. Yet, if queried about what a mathematician does, many would have some difficulty in replying much beyond the feeble, though correct, answer that a mathematician does mathematics. No, this is not an attempt to make a professional mathematician out of the reader! Not at all. Yet, it is felt to be important that a person should know what a mathematician does just as much as he should know what a doctor or architect or bus driver does.

It is true that a mathematician does mathematics. Nevertheless, some amplification is needed. But, first, a small digression is in order. Even though you may not become a skilled mathematician, each time you study mathematics you gain more insight into the process by which the subject developed. And mathematics is a subject learned by *doing*, because it involves a language that must be used to be learned, in much the same way as Italian or French or shorthand.

If you plan to learn anything from this text, you must study it with paper and pencil (and eraser!) at hand. The great majority of your mathematical knowledge will flow through the paper, the tip of the pencil, the fingers, and up the arm, and thence on into the head. Much of a mathematician's accomplishment can be measured in direct relation to the amount of crumpled paper in his wastebasket.

Some mathematicians create new mathematics and then pass their ideas along to others; other mathematicians use these ideas in other areas such as engineering, computer technology, actuarial work, business, etc. This section deals with those who create mathematics. Essentially, they try to create new systems or amplify and explain existing ones. As we have seen, a mathematical system consists of the logical development and proof of new statements from undefined terms, postulates, and definitions. The inexperienced student of mathematics may think that this is all that mathematics is about: the proof of theorems. This side of the subject is what is called formal mathematics; formal geometry is a part of it.

However, there is another side of the coin, and, for many, it is perhaps the more interesting one. This is the question of where do the theorems to be proven come from. For Euclid, his theorems came from the earlier practical work of the Egyptians, Babylonians, and early Greeks. For the present-day mathematician, his theorems come from his observations in the real world, his intuition and experimentation, and sometimes from pure guesswork and luck. The mathematician does not experiment in a random fashion however. Much of his experimentation is carried out with the full knowledge that the formal portion, and its necessary proofs, must follow. All such reasoning might be described as **inductive,** as opposed to the deductive reasoning of the formal side of mathematics. This mathematics is informal mathematics; in this book it will be informal geometry.

Recognize that the informal geometry must come first. There can be no theorems to prove if there are no theorems to be stated, or perhaps to use a better word, to be conjectured. Theorems usually state some relationship between or about geometric objects. These relationships are discovered by prior experimentation with specific cases and then generalized into a theorem covering each specific case in one statement.

It is the plan of this book to follow such a course of action. Informal or experimental geometry will come first to enable you to discover the existing relationships. Experimentation will involve manipulation and measurement of real objects for which the objects of our goemetry will be the mathematical models. At times, the real objects will be no more than drawings. Then, the formal aspect of geometry will be examined. It has a legitimate place in the study of the subject of geometry (as well as most other mathematics) and should not be ignored. A serious, rigorous development is not intended but rather enough of the so-called axiomatic method will be studied so that you can get a feeling for what it implies.

It is not the intent of this section to convey the notion that formal and informal geometry must be included at all levels of study of geometry. Certainly, the amount of formality is a function of age. Elementary school children should consider almost exclusively informal geometry, whereas the informal geometry must be included at all levels of study of geometry. of the type we are suggesting here. The serious mathematics student should have a great deal of the informal activity suggested for the mathematician; that is, he should get the thrill of formulating new theorems, even if they are new only to him.

EXERCISE SET 1.1

In some of these exercises and those that follow, it will be necessary to have a ruler, preferably subdivided into inches and centimeters, a protractor to measure angles, and a compass to draw circles. These exercises are designed to give you some experience in "experimentation" and conjecturing.

1. Draw a fairly large triangle on a piece of stiff cardboard. Use your ruler to locate the midpoint of each of the three sides. Draw the three medians (a median is a line segment from the vertex of the angle to the midpoint of the oposite side). These three segments should meet in a point. (It may not be exactly so with your drawing—why not?) Now cut out the triangle. Attach a piece of string or thread to a pin and then stick the pin into the point where the medians meet. Suspend the triangle by the thread. What do you observe?

2. Draw a four-sided figure (quadrilateral) so that none of the four sides crosses each other. Locate the midpoints of each of the four sides and then join the midpoints in order by straight line segments so that you form another quadrilateral with sides that do not cross each other. Repeat this with several other quadrilaterals. What seems to be true about the quadrilaterals formed by joining the midpoints?

3. Draw two line segments so that they intersect at a point O. Call these l_1 and l_2. On l_1, locate three distinct points P_1, P_2, and P_3 so that none of them coincides with point O. On l_2, locate three distinct points Q_1, Q_2, and Q_3, again so that none of the points coincides with point O. Draw the lines connecting P_1 with Q_2 and also P_2 with Q_1. Call the point of intersection of these two lines X. Do the same with P_2Q_3 and

P_3Q_2, calling this point of intersection Y. Finally do this again with P_1Q_2 and P_3Q_1, calling this point Z. What do you observe about points X, Y, and Z?

4. Is the location of points X, Y, and Z just an accident in Problem 3? Repeat the exercise at least twice more to convince yourself about the answer.

5. If you are like most persons, you probably placed point P_2 between points P_1 and P_3, and Q_2 between Q_1 and Q_3. Repeat the above "experiment" one more time with a variation in the order of placing the points on the lines and see if the phenomenon still persists.

6. If you did Problem 5, you may have had the following variation of this experiment come to mind. It is quite possible to place points P_1, P_2, and Q_1, Q_2 so that lines P_1Q_2 and P_2Q_1 are parallel—that is, they will not intersect and there will be no point X. If you have not tried this variation, do it now. Before you complete the experiment by drawing the two remaining pairs of lines, make a guess about the relative positions of the points Y and Z.

7. Do you think that the original stipulation of l_1 and l_2 meeting at point O is a necessary ingredient of this "experiment"? Will the same results occur if l_1 and l_2 are parallel?

8. With your compass, draw a circle and locate on it six random points, named P_1, P_2, P_3, P_4, P_5, and P_6. Find the points of intersection of the pairs of lines determined by the following pairings of points: P_1P_2 with P_4P_5, P_2P_3 with P_5P_6, and P_3P_4 with P_6P_1. If you are lucky, these three points of intersection all "landed" on your paper. What do you observe about these three points? Repeat this exercise at least twice more.

9. From a point O draw three distinct lines, l_1, l_2, and l_3. Now draw two distinct triangles, ABC and $A'B'C'$, so that each triangle has one of its vertices on one of the three lines—that is, vertices A and A' should lie on line l_1, B and B' on line l_2, and C and C' on l_3. Now find the points of intersection of the lines determined by the corresponding sides of the two triangles—that is, the points of intersection of AB with $A'B'$, AC with $A'C'$, and BC with $B'C'$. What do you observe about these three points?

10. Draw an equilateral triangle (one with all three sides equal in length). Using each side of the triangle as a base, construct a square upon each of the three sides. Now locate the point of intersection of the two

diagonals for each of these three squares. Connect these three points to form a triangle. What kind of triangle does it appear to be?

SUGGESTED READINGS

Complete bibliographical information is given in the bibliography at the end of the text. Numbers in brackets refer to the number preceding the name of the text as listed in the bibliography.

[7]

[11], Ch. 1

[16]

[20], Ch. 1

[29], Ch. 1

2

SETS, POINTS, LINES, PLANES AND SPACE

The purposes of this chapter are twofold. First, a review of the basic notions of sets will be given. It is assumed that the reader has had prior experience with these ideas in an earlier course in which the development of the real number system has been studied. This time the ideas will be restated in terms of geometric considerations. Secondly, some of the basic terminology of geometry will be introduced. Whether considering the formal or informal aspect of the subject, it will be impossible to proceed very far without some common terms whose meanings are generally agreed upon. Young children are not renowned for their conversational ability, mostly because they lack the vocabulary with which to communicate; one is similarly handicapped in geometry until he learns the mathematical vocabulary to communicate geometric ideas.

2.1 SETS AND SUBSETS

One of the most basic notions in all of mathematics is the concept of set.

9

A definition of a set could be attempted but the definition would, of necessity, require the usage of a word or words that are synonymous with the word "set." One could consult a dictionary to seek the meaning of the synonymous words, but eventually he would come full circle to the place where the word "set" was used in one of the definitions. Thus, we shall not attempt to make a definition but agree to use it as an undefined term. This agreement is no disadvantage, since the meaning of "set" is generally agreed upon for common usage, and it can be described by means of some of the synonyms mentioned below. Hopefully, and such is usually the case, one of the words used will already have meaning for the reader.

A **set** then, is described as a collection, or aggregate, or group of objects. These objects may be real, such as people, or they may be abstractions, such as numbers. Other words that might have been used are herd, covey, flock, class, or school, but these have somewhat specialized connotations connected with certain types of objects and thus might be misleading. A set may be a collection of any objects with nothing more in common than the fact that they are elements of the same set. The objects in the set are usually called **elements** or **members** of the set. It is common to denote a set symbolically by use of a capital letter; for example,

$$D = \{ \text{Monday, Tuesday, Wednesday, Thursday, Friday} \}$$

The elements of a set, when referred to in a generalized fashion, are denoted by small letters. We will write $a \in A$ to mean "a is an element of set A," or "a is a member of set A." The notation, $a \notin A$, is used to denote that a is *not* an element of set A.

Several items should be noted. First, while a set may be a collection of seemingly disassociated objects, the thing that is important is that the set must be described or defined in such a way that it is perfectly clear whether or not any given object is a member of the set. Secondly, there is nothing in the discussion to preclude the objects in a set from being sets themselves. For example, the set of all teams in the National Football League consists of those elements that happen to be sets themselves, teams of individual players. Likewise, the set of all right triangles consists of entities that, as we shall presently see, may be considered as sets of other entities known as points.

Related to the idea of a set is the idea of a subset of a given set. It is defined as follows:

DEFINITION 2.1

A set A is a **subset** of a given set B if and only if every element of A is also an element of B.

The symbolic notation for the relation "is a subset of" is $A \subseteq B$, which is read "A is a subset of B" or "set B contains set A."

EXAMPLES

(a) If T is the set of all triangles and R is the set of all right triangles, then $R \subseteq T$.

(b) If P is the set of all polygons and T is the set of all triangles, then $T \subseteq P$.
(If you don't know what a polygon is, don't despair! It will be explained in Chapter 5.)

(c) The set E of even numbers is subset of the set I, the set of integers; thus $E \subseteq I$.

Note that the definition allows the possibility of a set being a subset of itself; this fact is true for *all* sets. Thus, if we were being formal here, this fact could be stated as a theorem, the proof of which would follow directly from Definition 2.1.

One set in particular needs special mention. It is the set that has no elements in it. There are many ways of describing this same set, and for this reason it is called the **empty set** (or the **null set**) and is symbolized by $\varnothing$. (One of its many descriptions might be the set of all living female ex-Presidents of the United States.) It turns out to be a fact that the empty set is a subset of all sets. This fact, also, could be formalized as a theorem, since it is always true. Its proof follows from Definition 2.1 in a way that is not as obvious as the fact mentioned in the previous paragraph. Although the first portion of this text is not given over to formal geometry, let us see how we could be formal in an "informal" way. We mean that we shall not write out a proof in a formal manner; rather, it will be explained in a reasonable, conversational manner.

Definition 2.1 says $A \subseteq B$ if and only if *every* element of A is also an element of B. In a manner of speaking, we can say this is true in the case of $\varnothing$, since every element of $\varnothing$ *is* in B. Of course, it just so happens that $\varnothing$ does not have any elements! Now, this line of reasoning may be objectionable to many persons and rightly so. But it does give a hint of another method of reasoning that gets at the problem in an indirect way. Suppose $\emptyset$ is *not* a subset of B. Then, this would mean that the definition of subset would be violated in some way. The only way that the definition's conditions cannot be met is for A (of the definition—it is $\emptyset$ in this particular case) to have an element in it that is not in B. Now this is plainly impossible when A is the empty set, because $\emptyset$ contains no elements and thus cannot have an element not in B. Thus, the definition of subset has *not* been violated in the case of $\emptyset$, and since we have no other possibilities, it must, of necessity, be a subset of B.

EXERCISE SET 2.1

1. For the following sets, determine whether or not they are **"well-defined"** in the sense that it is possible to tell with certainty whether or not a given element is a member of the set.

   (a) The set of prime numbers less than 100.
   (b) The set of good mathematics teachers you have had.
   (c) The set of female heads of state.
   (d) The set of four-sided geometric figures.
   (e) The set of good-looking co-eds on your campus.
   (f) The blades of grass in the world.
   (g) The set of future presidents of the United States.

2. For each of the following sets, list all the subsets.
   (a) $A = \{a\}$
   (b) $F = \{\triangle, \square\}$
   (c) $T = \{m, n, p\}$
   (d) $B = \{w, x, y, z\}$

3. (a) Count the number of subsets for each of the parts of Exercise 2.
   (b) On the basis of your results, make a conjecture as to the number of subsets a set with five elements would have. With six elements.

4. On the basis of Exercises 2 and 3, make a conjecture as to the number of subsets a set of n elements would have, where n is any counting number. Try to write out an "informal" proof of your conjecture. It is important that you write it out; doing so will force you to organize your thinking.

5. Write out five different descriptions of the empty set. Do not be afraid to let your imagination and sense of humor be your guide.

6. Consider T as the set of all triangles. Describe five different sets that

are subsets of T. Are any of the subsets mentioned subsets of each other? Do any of the subsets overlap in any way?

7. Repeat Exercise 6, using Q as the set of all quadrilaterals (four-sided figures).

2.2 POINTS AND SPACE

The fundamental notion in the study of geometry is that of a **point**. Most persons have a fairly good idea what a point is. Euclid called it "that which has no part." Since it is a basic idea, we shall leave the word "point" as undefined. The thing that is important to remember is that the notion of point is an abstraction, which is a mathematical model of something real. That real thing is perhaps the end of a pencil, or the tip of your nose, or the intersection of two roads on a map. The picture used to represent a point is a dot. Since a point has no dimensions, the smaller the dot is, the better it represents the point. Of course, it is by no means perfect, since the best representation would have to be the dot that we could not even see. However, to be useful as a picture, the dot must be large enough for the eye to locate. A point is labeled with a capital letter, as shown in Fig. 2-1.

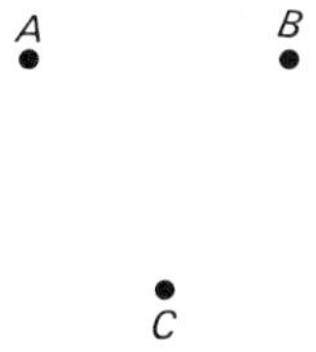

$$A \qquad\qquad B$$
$$\bullet \qquad\qquad \bullet$$

$$\bullet$$
$$C$$

Figure 2-1

In much the same fashion, most people have a strong intuitive feeling for what is meant by the term **space**. It could very easily be taken as an undefined term, but it shall not because the word may be defined very simply in terms of words and ideas that have already been discussed and whose meanings have been agreed upon.

DEFINITION 2.2

Space is the set of all points.

Studying only the set of all points would be somewhat boring. The interest lies in particular subsets of this set, especially those that are so located as to form certain familiar geometric figures that are worthy of study. Any set of points is given the general name of **geometric figure**. It is felt there will be no confusion between this use of the term "figure" and the other use, which shall be the reference to the various pictorial illustrations placed throughout the text.

2.3 UNION AND INTERSECTION

The operations of set union and set intersection should be familiar to the student and are given here for the sake of completeness. Given two sets, the binary operation of set union may be performed on these two sets.

DEFINITION 2.3

The **union** of two sets, A and B, is a set whose elements are the totality of elements in A and in B.

The union of two sets A and B is denoted by $A \cup B$.   Essentially, if an element is in $A \cup B$, then that element is an element of A or an element of B or an element of both A and B. In mathematics, this fact is stated by saying that if an element is in $A \cup B$, then it is in set A or B. This usage of the word "or" is different from the common usage of "or" that carries an exclusive meaning in the sense of "either . . . or." The mathematical "or" is thus called **inclusive** rather than exclusive.

This fact can be represented pictorially as in Fig. 2-2. Various relation-

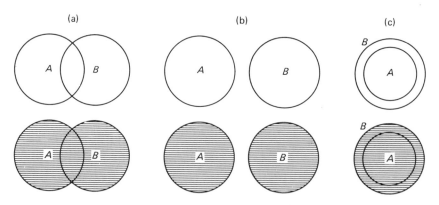

Figure 2-2

ships between sets A and B are pictured, and then $A \cup B$ is shown by the horizontal shading.

The second binary operation that may be performed on two sets is that of set intersection.

DEFINITION 2.4

The **intersection** of two sets, A and B, is a set whose elements are all those elements that are shared commonly by A and B.

The intersection of two sets A and B is denoted by $A \cap B$. If an element is in $A \cap B$, then that element is in both sets and we say that element is in set A *and* set B. If the intersection of two sets happens to be the empty set, then those sets have a special name.

DEFINITION 2.5

Two sets, A and B, are said to be **disjoint** if they have no elements in common; that is, if $A \cap B = \varnothing$.

Figure 2-3 illustrates the concept of intersection that is pictured by the horizontal shading. Notice that part (b) of Fig. 2-3 has no shading, since the sets A and B as pictured are disjoint. Consequently, their intersection is the empty set.

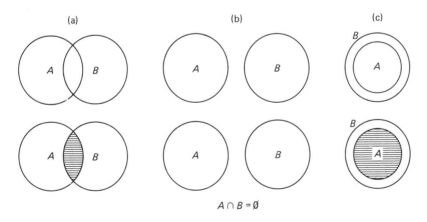

$$A \cap B = \varnothing$$

Figure 2-3

It is well to remember that these two operations of set union and set intersection have the properties of being both commutative and associative.

The **commutative** properties are:

$$A \cup B = B \cup A$$
$$A \cap B = B \cap A$$

The **associative** properties are:

$$(A \cup B) \cup C = A \cup (B \cup C)$$
$$(A \cap B) \cap C = A \cap (B \cap C)$$

EXERCISE SET 2.2

1. Complete each of the following to make a true statement.

   (a) $A \cap \varnothing = $ _____ .   (c) $A \cap A = $ _____ .

   (b) $A \cup \varnothing = $ _____ .   (d) $A \cup A = $ _____ .

   (e) If $A \subseteq B$, then $A \cup B = $ _____ .

   (f) If $A \subseteq B$, then $A \cap B = $ _____ .

   (g) If $B \subseteq A$, then $A \cup B = $ _____ .

   (h) If $B \subseteq A$, then $A \cap B = $ _____ .

2. Find the intersection for each of the following:

   (a) $A = \{a, b, c\}$, $B = \{c, d, e\}$

   (b) $P = \{W, X, Y\}$, $Q = \{X, Y, Z\}$

   (c) $E = \{0, 2, 4, 6\}$, $O = \{1, 3, 5\}$

   (d) $M = \{1, 2, 3, 4\}$, $N = \{4, 3, 2, 1\}$

   (e) $X = \{1, 2, 3, 4\}$, $Y = \{1, 2\}$

   (f) $X = \{a, b, c, d\}$, $Y = \{b, c, d\}$, $Z = \{d, e, f\}$

   (g) $R = \{3, 6, 9\}$, $S = \{2, 4, 6\}$, $T = \{2, 3, 5, 7\}$

3. Find the union in each of (a) through (g) in Exercise 2.

4. Draw Venn diagrams, such as those in Fig. 2-2 and 2-3, and use appropriate shading to illustrate the following distributive laws.

   (a) $A \cup (B \cap C) = (A \cup B) \cap (A \cup C)$

   (b) $A \cap (B \cup C) = (A \cap B) \cup (A \cap C)$

   (*Hint*: For each part draw the diagram twice, once to illustrate the left-hand side of the equal sign and once to illustrate the right.)

2.4 LINES, SEGMENTS, AND RAYS

A particular set of points that is of great importance in the study of geometry is that which is described as a **line**. We shall qualify it further by restricting our discussions, unless otherwise stated, to **straight lines**. What is meant by a "straight line"? Should the often-heard definition of "a straight line is the shortest distance between two points" be used? Preferably not, for two reasons. First, we would like to think of a line as extending infinitely far in both directions, and this idea does not fit with that of "shortest distance between two points." Secondly, when we deal with sets of points, the concept of distance does not have too much meaning at this time.

Therefore, the word "line" is taken as another undefined term with the assumption that "line" and particularly "straight line" have the same meaning for all readers. While this seems to be a fair assumption, it might be wise to point out that not always is it true that everyone has the same intuititive notion for certain words. Such is apt to be the case with young children whose ideas are not yet completely formed and molded to the adult way of thinking. For example, if we were to ask a child to look at Fig. 2-4

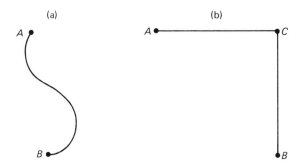

Figure 2-4

and then ask him whether part (a) or part (b) is a picture of a straight line, he might very well answer "(a)," because in it the line goes directly from A to B without any detours via a location that has been labeled as C in part (b). This response may come from frequent admonitions from Mother to "come straight home from school."

An etymological examination of the words "straight line" is revealing. They come from words that originally meant "stretched string." It is not hard to see the connection between "straight" and "stretched"; it is also easy to see the connection between the other two if we look at the word "line," which we see is similar to "linen," which at one time had a meaning of "string."

A line is symbolized in one of two ways. One way is to use a small letter, such as l or m. The other way is to choose two distinct points on the line, say A and B, and write $\overleftrightarrow{AB}$ to stand for the line that contains these two points. The significance of the two arrowheads is the reminder that a line is to be thought of as going infinitely far in both directions. When we draw a figure that represents a line it shall be indicated as shown in Fig. 2-5. The simpler phrase, "draw a line," will be normally used, even though it is technically incorrect.

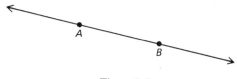

Figure 2-5

Many times, rather than use a line, it is preferable to consider what is known as a line segment, or more simply, a segment.

DEFINITION 2.6

A subset of a line is called a **line segment** if it consists of two distinct points on that line and all points of the line between those two points.

A segment is denoted by $\overline{AB}$, where A and B are the two points referred to in the definition. Points A and B are called the endpoints of the segment. It must be mentioned that Definition 2.6 includes a word that has not been previously defined or discussed. This is the word "between." One of the greatest complaints present-day mathematicians have of the geometry of Euclid is that he did not make a clear distinction as to what he meant by the word "between" or by the notion of "betweenness." Although a more rigorous and formal treatment of geometry can make this word quite clear and properly defined, we shall, at this time, take it as an undefined word whose meaning is the same for everyone. The only restriction that will be imposed is that we shall talk of betweenness of points only when *they are on the same line*. It is almost obvious to say that if we consider three distinct points, A, B, and C, on a straight line, one and only one of the three is between the other two. Figure 2-6 illustrates that this is not necessarily so if another kind of curve is considered.

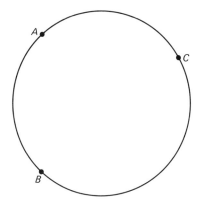

Figure 2-6

We would like to introduce one more term, that of ray. To do this properly another term must be first introduced. Consider a line and a point on that line. This point partitions the line into three distinct, mutually disjoint subsets. One of the subsets is that singleton set consisting of the point itself; the other two are what are called **half-lines** and can be thought of as two infinite sets of points, one set being on one side of the point and the other set on the other side of the point. The partitioning point is called the endpoint of either half-line, although it is not an element of either half-line. This situation is analogous to taking the number "0" and considering it as partitioning the set of integers into three disjoint sets: $\{0\}$ the set of positive integers, and the set of negative integers. This is not too surprising an analogy when one thinks of the integers being placed on the familiar number line.

We are now ready to define ray.

DEFINITION 2.7

A **ray** is a set of points consisting of the union of a half-line and its endpoint.

A **ray** is designated by naming its endpoint and one other point on the half-line and using the notation $\overrightarrow{AB}$, with the convention that the letter listed first is the name of the endpoint. The situation is pictured in Fig. 2-7. Given line l with points, A, B, and C, as pictured, if B is considered as the

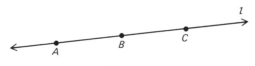

Figure 2-7

partitioning point, then we have two rays, $\overrightarrow{BC}$ and $\overrightarrow{BA}$, which are often called opposite rays.

Be alert to the fact that $\overrightarrow{AB} \neq \overrightarrow{BA}$.   It might be argued that, in order to promote accuracy and ease of understanding, $\overrightarrow{BA}$ might better be written as $\overleftarrow{AB}$, since then the arrowhead would be the deciding factor. If arrowheads were changed, then the convention of the first letter listed being the endpoint would have to be discarded. Moreover, such an argument assumes that all rays lie along horizontal lines, which is certainly not the case. If we demanded that the arrowhead point in the direction of the ray, we would have a difficult time symbolizing vertical and oblique rays. Thus, we adopt the convention $\overrightarrow{AB}$ regardless of the direction of the ray; this convention simply tells us that A is the endpoint.

Before leaving this section, we should mention that a definition of ray can be given without reference to the concept of half-line. We might have said

$$\overrightarrow{AB} = \overline{AB} \cup \{ \text{ all points } C \text{ on } \overleftrightarrow{AB} \text{ such that } B \text{ is between } A \text{ and } C\}$$

This definition seems a little more formal and less intuitive and thus was not chosen even though it is perfectly correct. It does point up, however, a fact that is often true in mathematics, namely that there is more than one way to make a valid definition.

EXERCISE SET 2.3

1. Consider the line shown below with its marked points; then fill in the blanks.

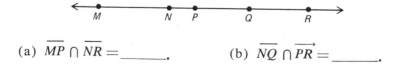

(a) $\overline{MP} \cap \overline{NR} = $_____.   (b) $\overrightarrow{NQ} \cap \overrightarrow{PR} = $_____.

(c) $\overline{PQ} \cup \overrightarrow{NR} =$ _____.   (f) $\overline{MN} \cap \overline{RQ} =$ _____.

(d) $\overrightarrow{NM} \cap \overrightarrow{PQ} =$ _____.   (g) $\overrightarrow{MQ} \cup \overrightarrow{NP} =$ _____.

(e) $\overline{NP} \cap \overline{PQ} =$ _____.   (h) $\overrightarrow{PN} \cup \overrightarrow{PQ} =$ _____.

2. Draw a vertical line. Label four distinct points, A, B, C, D, in that order from bottom to top. Name two segments

   (a) whose intersection is a point;
   (b) whose intersection is empty;
   (c) whose intersection is a segment;
   (d) whose union is a segment;
   (e) whose union is not a segment.

3. Repeat Exercise 2, except name two rays.

4. Draw two segments $\overline{MN}$ and $\overline{PQ}$ for which $\overline{MN} \cap \overline{PQ}$ is empty but $\overleftrightarrow{MN} \cap \overleftrightarrow{PQ}$ is one point.

5. Draw a line m and on it mark points X, Y, and Z.

   (a) How many distinct segments have been formed by these points? Name them.
   (b) How many distinct rays have been formed by these points? Name them.
   (c) How many distinct half-lines have been formed by these points? Name them.

6. Repeat Exercise 5, parts (a), (b), and (c), with line m and four distinct points, W, X, Y, and Z.

7. From your work in Exercises 5 and 6, make a conjecture as to the answers to (a), (b), and (c) if there were five distinct points; six points; n points.

8. On your paper locate three points, D, E, and F, so that all three are not on the same line. How many different segments can be drawn?

9. Repeat Exercise 8 with four distinct points, D, E, F, and G, so that no three are on the same line. How many different segments can be drawn?

10. On the basis of Exercises 8 and 9, make a conjecture as to the number of different segments that could be drawn if there were five distinct points, no three of which are on the same line; six points; n points.

11. Distinguish between a ray and a half-line.

12. Formulate a definition for $\overrightarrow{BA}$ similar to the one in the last paragraph of Section 2.4.

2.5   PLANES

The final geometric figure of interest at this stage of the development is the **plane**. As a set of points, it is an abstraction. Yet, it is a good mathematical model of a class of real objects that all possess the property of flatness, for example, the top of a desk, the surface of a blackboard, sheets of paper. Although it is an abstraction, the plane is often described as a two-dimensional object having only length and width (no thickness), these two dimensions stretching infinitely far in the same sense that a line extends infinitely far in its one "dimension." In fact, the "straightness" of the line and the "flatness" of the plane are analogous properties that lead to the statement that if two points of a line are on a plane, then all points of the line are on the plane. (Notice the usage of the terminology "on a plane" in the preceding sentence; we speak of points as lying *on* a plane.)

This description of a plane in no way should be construed as a definition of a plane. The term plane is taken as an undefined term, along with point and line.

In depicting a plane, we find it impossible to draw an entire plane for the same reason it is impossible to draw an entire line. Thus, only a portion of it can be shown, and, as exemplified in Fig. 2-8, is done by drawing parallelograms. The easiest way to talk of a plane is to give it a name; in this text, a plane is named by a small Greek letter, for example, plane α.

Now consider a line lying in a plane. In a manner analogous to a point lying on a line, the line partitions the plane into three disjoint subsets, one being the line itself, and the other two being sets of points called **half-planes.** The line is called the **edge** or **boundary** of the half-planes; it is not part of either half-plane. If one wishes to consider the union of a half-plane and its boundary, the term **closed half-plane** is used. A closed half-plane is the analogue of the ray. Sometimes the term **open half-plane** is used instead of the simpler "half-plane" in order to distinguish it from the closed half-plane.

The points in a half-plane are sometimes referred to as being "on the same side of" the line that is the boundary. In order to name half-planes, the customary procedure is to choose a point in one of the half-planes (remember the definition states that this point is not on the boundary line),

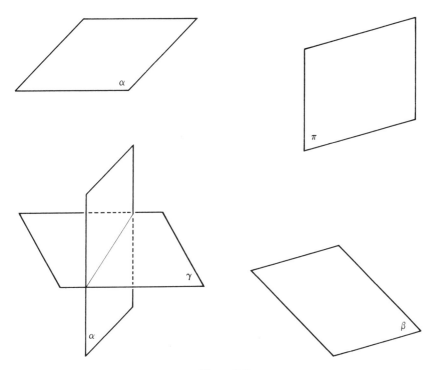

Figure 2-8

say the point P, and call the half-plane the "P-side" of the line. A picture of this situation is shown in Fig. 2-9.

This discussion can be continued one step further (it might be better to say "one dimension further"). In precisely the same manner as the

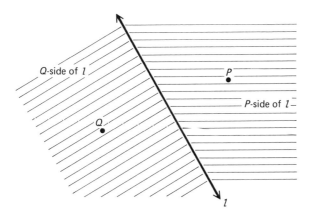

Figure 2-9

previous two instances, it is possible to say that a plane partitions space into three distinct and disjoint subsets, the plane itself and two other sets called **half-spaces**. The plane that separates the two half-spaces is called the **boundary** of the half-spaces. A physical representation of this concept might be a wall that separates space into two parts, each on separate sides of the wall. As before, if the union of a half-space and its boundary is considered, the term **closed half-space** is used to name this union.

2.6 CONVEX SETS

While the word "convex" may have meaning for many readers, the term "convex set" is probably new to most. Perhaps an easy way to describe something that is convex is to say that it has no dents in it. While such a description is far from one that we would like to use with mathematical abstractions known as sets, it does give a hint. Before we give the definition of convex sets, several pictures of various sets of points are given in Fig. 2-10.

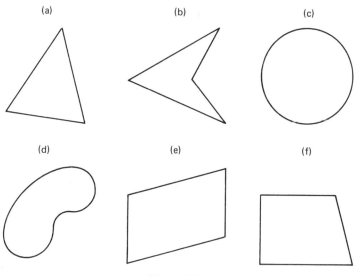

Figure 2-10

Consider the "interior" of the figures to be part of the set. Try to decide before reading further which pictures represent convex sets. Recognizing that these figures are only specialized kinds of geometric figures (as we shall see later, they are all called "simple closed curves"), you hopefully came up with the correct answers of (a), (c), (e), and (f) as being convex.

Questions about the convexity of other kinds of sets of points will be asked in the exercises. After reading the following definition, go back to Fig. 2-10 and try to see how the definition fits the special cases we have shown.

DEFINITION 2.8

A set S is a **convex set** if and only if, for any two distinct points A and B in S, all points of the segment $\overline{AB}$ are also in S. It is also agreed that a set consisting of only one point, and the empty set, shall be considered convex sets.

Even though the pictures in Fig. 2-10 are representatives of figures in the plane, Definition 2.8 is broad enough to cover cases of three-dimensional figures.

2.7 SET EQUALITY

Since we shall be dealing with sets of points throughout our study of geometry, it would be well to review precisely what is meant by equality of sets in particular and equality in general. We have the knowledge to give an exact definition of set equality.

DEFINITION 2.9

For any two sets A and B, $A = B$ if and only if $A \subseteq B$ and $B \subseteq A$.

In everyday language, to say that two sets are equal is equivalent to saying they have exactly the same elements. Geometrically speaking, we may not use the symbol "$=$" unless the two things being equated have exactly the same points. For example, in considering the isosceles triangle pictured in Fig. 2-11, one might be tempted to write $\overline{AB} = \overline{AC}$. $\overline{AB}$ and $\overline{AC}$ are obviously not the same set of points and so the symbol "$=$" should not be used between them. What was meant was that $\overline{AB}$ and $\overline{AC}$ were the same length. The idea of measurement and length will be studied in greater detail in a subsequent chapter in which it will be seen that the correct term to use with respect to $\overline{AB}$ and $\overline{AC}$ is "congruent."

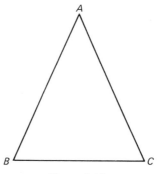

Figure 2-11

When the symbol "$=$" is used with respect to numbers, the accepted convention is that the right and left members will be representations or symbols for the same number, or, as it is often said, they will be two names for the same number.   When an equation such as $3x + 5 = 13$ is written, the "$=$" sign is usually taken to mean "give the replacement or replacements for the variable x such that the left and right sides will represent the same number or numbers."   In some instances, equality will hold for all possible replacements for x for which the expression makes sense (in which case we have what is known as an **identity**); for others only some or no replacements will work (then it is usually called a **conditional** equation).

2.8   ONE-TO-ONE CORRESPONDENCE

A very useful concept when working with sets is that of a one-to-one correspondence.   Its definition is given.

>   **DEFINITION 2.10**
>
>   Two sets, A and B, are said to be in a **one-to-one correspondence** if and only if for each element in A there corresponds exactly one element of B, and for each element of B there corresponds exactly one element of A.

Several examples will be given to clarify the idea.

EXAMPLES

(a) $A = \{\, a, b, c \,\}$ $\qquad\qquad$ $B = \{\, 1, 2, 3 \,\}$

$$\{\, a, b, c \,\}$$

$$\updownarrow \;\; \updownarrow \;\; \updownarrow$$

$$\{\, 1, 2, 3 \,\}$$

Here the matching is indicated by the double-headed arrows. This is not the only way in which the matching could have been arranged. Shown below is another possibility.

$$\{\, a, b, c \,\}$$

$$\times \;\; \updownarrow$$

$$\{\, 1, 2, 3 \,\}$$

(b) $N = \{\, 1, 2, 3, 4, 5, \ldots \,\}$

$$\updownarrow \; \updownarrow \; \updownarrow \; \updownarrow \; \updownarrow$$

$S = \{\, 1, 4, 9, 16, 25, \ldots \,\}$

(c) The set of men and the set of women who get married in a given city on a given day. Here the manner of matching is obvious.

(d) An example that appears in mathematics courses is the one-to-one correspondence between points on a number line and the set of real numbers. The student who needs to review the real numbers is referred to Section 4.2 of this text.

(e) There is a one-to-one correspondence between the sets of points that are the two circles pictured. The rays whose endpoints are O show one way of setting up the matching of points.

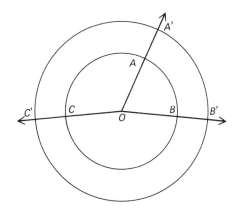

(f) This is somewhat similar to example (e).   Some of the matchings between points of $\overline{AB}$ and $\overline{MN}$ are shown.

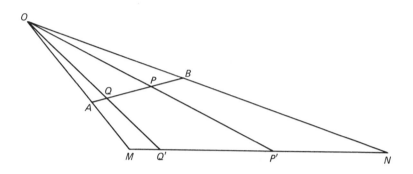

2.9 EQUIVALENCE RELATIONS

One-to-one correspondence is a relationship that exists between two sets. As such, it is an example of a relationship known as an equivalence relation. Let us take a few lines to review this important concept.   First, let us try to make clear the idea of a relation defined on a set.

DEFINITION 2.11

A **relation** R is defined on some set S if for *all* ordered pairs (a, b), where a and b are elements of S, the statement "aRb" makes sense and is either true or false.

This is a rather abstract definition; perhaps some examples will make it clear.

EXAMPLES

(a) Let S be the set of natural numbers.
Let R be the relation "is the square of."
Then, for the arbitrary selection of the pair $(9, 3)$, the statement "$9R3$" can be translated into the English sentence: "9 is the square of 3."   This sentence makes sense and is also true.
"$9R4$" means "9 is the square of 4"; as such it makes sense but it is false.

The relation "is a square of" is a well-defined relation on the set of all natural numbers.

(b) Let S be the set of all people.

Let R be the relation "lives on the same street as."

Then, statements such as "Mary lives on the same street as Lucille" are meaningful ones whose truth can be determined. Therefore, "lives on the same street as" is a relation on the set of all people.

(c) Let S be the set of all triangles.

Let R be the relation "has the same area as."

Statements such as "$\triangle ABC$ has the same area as $\triangle DEF$" are meaningful ones.

(d) "is a factor of" is a relation defined on the set of all integers.

(e) "won more games than" is a relation defined on the set of all major league baseball teams.

(f) "has more money than" is *not* a relation defined on the set of all line segments.

(g) "is in one-to-one correspondence with" is a relation defined on the set of all sets.

When the ordered pair (a, b) makes the statement "aRb" a true statement, it is often said that "a is related to b."

As a special class of relations, we have those that are called equivalence relations. They are defined as follows:

DEFINITION 2.12

A relation defined on a set S is an **equivalence relation** if and only if it has the folowing three properties:

(1)   reflexive: For *all* $x \in S$, xRx;

(2)   symmetric: For *all* $x, y \in S$, if xRy, then yRx;

(3)   transitive: For *all* $x, y, z \in S$, if xRy, yRz, then xRz.

EXAMPLES

(a) The relation "lives on the same street as" is certainly reflexive, since, for any person, he lives on the same street as himself.

(b) The relation "won more games than" is not reflexive. For example, it is false to say "Detroit won more games than Detroit." However, it is a transitive relation.

(c) The relation "is a factor of" is not symmetric; as a counter-example we have the true statement "2 is a factor of 8" but the false statement "8 is a factor of 2."

If a relation defined on a set is an equivalence relation, it has the property of partitioning that set into smaller subsets called **equivalence classes**, consisting only of those elements that are related to each other. Each of these equivalence classes is disjoint with every other one, but they do have the property that the union of all of them is equal to the original set.

EXAMPLES

(a) Consider the set of natural numbers and the relation "has the same remainder as ____ when divided by 6." This relation partitions the natural numbers into six equivalence classes:

$$[1] = \{1, 7, 13, 19, 25, \ldots\}$$
$$[2] = \{2, 8, 14, 20, 26, \ldots\}$$
$$[3] = \{3, 9, 15, 21, 27, \ldots\}$$
$$[4] = \{4, 10, 16, 22, 28, \ldots\}$$
$$[5] = \{5, 11, 17, 23, 29, \ldots\}$$
$$[6] = \{6, 12, 18, 24, 30, \ldots\}$$

(b) Consider the set of students in a given senior high school and the relation "is in the same grade as." This relation partitions the student body into three equivalence classes: sophomores, juniors, and seniors.

EXERCISE SET 2.4

1. Decide whether each of the following is a convex set. Draw a picture if it helps you.

(a) A segment
(b) A ray
(c) A line
(d) A half-line
(e) A half-plane
(f) A half-space
(g) Union of two segments

(h) Union of two rays
(i) Intersection of two segments
(j) Intersection of two rays
(k) Union of two convex sets
(l) Intersection of two convex sets

2. Formulate a manner of testing quite precisely whether two points are in the same half-plane.

3. For the two sets A and B of example (a) of Section 2.8, in how many different ways could a one-to-one correspondence between these two sets be established? What would your answer be if A and B each had four elements? five elements? n elements? If A had four and B had five? If B had four and A had five?

4. Indicate in a manner similar to examples (e) and (f) of this section how a one-to-one correspondence would be set up between the points of a line segment and a circle.

5. Do the same for the two segments pictured below.

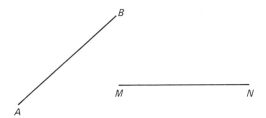

6. For each of the following sets and relations, decide whether the relation is one that can be defined as a relation on that set. If the answer is in the affirmative, then decide whether the relation is an Equivalence Relation.

Set	*Relation*
(a) Integers	is a divisor of
(b) Rationals	is relatively prime to
(c) People	is the father of
(d) Natural numbers	gives the same remainder as _____ when divided by 5
(e) Triangles	has the same shape as
(f) Line segments	has the same length as
(g) Lines	is parallel to
(h) Points	lies in the same half-plane as
(i) People	lies in the same half-plane as
(j) People	is a cousin of
(k) Points	lies in the same place as
(l) Integers	is between

7. A single line partitions a plane into two convex half-planes, where the half-planes are two-dimensional objects. What is the *maximum* number

of two-dimensional objects determined by two lines in the same plane? How about three lines in the same plane? Four?

8. This is much the same question as in Exercise 7, except that we shall inquire about planes partitioning space into three-dimensional objects. What is the *maximum* number of three-dimensional objects determined by two planes? By three planes? Four? Would you like to try for five?

SUGGESTED READINGS

[8], Ch. 4

[9], Ch. 2

[15], Ch. 3

[24], Ch. 2

[27], Ch. 1

3

INCIDENCE GEOMETRY

Thus far, we have reviewed some of the fundamental ideas of sets and have related these ideas to the particular class of sets of points. In addition to coming to some agreement on what is meant by the words "point," "line," "plane," and "space," we have briefly given some consideration to the notions of "separation" and "betweenness." These latter two ideas are parts of a main subdivision of the study of geometry known as **nonmetric geometry** which deals with relationships between sets of points in a context free of the idea of measurement or size. These relationships hold, regardless of the measure of the set of points. Measure or size pertains to angles and line segments as well as to other figures not yet mentioned but which can be thought of as having area or volume. The concept of measurement will be examined in the following chapter.

Also coming under the general heading of nonmetric geometry is the subject of the present chapter, **incidence geometry.** Incidence geometry has to do with the results of unions and intersections of points, lines and planes. With the exception of the section on angles, most of the material in this chapter will deal with the intersection of these sets of points.

Before beginning, let us first clear up one detail of language that may seem a bit unusual upon first reading but that proves to be very useful in the study of incidence geometry. Suppose that we have a given line *l*, one of whose points is the point A. Using set notation, this fact may be written $A \in l$. (Notice that this notation looks a little different from the usual way of writing that some element is a member of a set. The roles of the small and capital letters are reversed. However, this form must be used since the notation for points and lines that we have chosen forces us to do so.) We would say "A lies on line l"; we shall also say "line l lies on point A." It is this latter part that may be somewhat unfamiliar to the reader. In the same fashion, if we have a line $\overleftrightarrow{AB}$ contained in a plane α, we shall write "line $\overleftrightarrow{AB}$ lies on plane α" and also "plane α lies on line $\overleftrightarrow{AB}$." The set notation for this relation will be $\overleftrightarrow{AB} \subseteq \alpha$.

3.1 LINES AND POINTS

In this section and those that follow, we shall continue our study of geometry from the informal point of view. It may be pointed out that many textbooks and teachers of mathematics choose to begin their study of formal geometry with this material on incidence. The reasons for this choice are that the ideas dealt with (points, lines, and planes) are relatively simple ones. The theorems are statements that are intuitively clear and their proofs are simple to follow. However, the intuitive obviousness of the incidence relationships also makes possible a more informal approach.

A line is a set of points; although no formal statement has been made of the fact, it is accepted as true that there are infinitely many points lying on a line. If the roles of line and point are interchanged, then the question arises: How many lines are there lying on a point? It is also accepted as true that there are infinitely many lines lying on a point. An illustration of this fact is hard to draw—let us consider a "mental picture" prompted by the following physical model. Consider a pin cushion filled with pins sticking out in all directions. The pin cushion is the model of the point (albeit a rather large one!) and the pins may be thought of as the physical models for some of the infinitely many lines.

All of the lines that lie on a given point are given a special name.

DEFINITION 3.1

Two or more lines are said to be **concurrent** if and only if they lie on the same point.

Points and lines are fixed in space and, as such, cannot move. However, it is often convenient to think of them as movable in order to more easily visualize certain concepts. Such is the case in what follows.

Imagine a line as rotating about some fixed point in a manner similar to the minute hand of a clock. Imagine further placing the point of a pencil (this happens to be a good physical model for a point) in the path of the rotating line so that when the line came to the position of the point at the end of the pencil, it would be no longer be free to rotate and would become a fixed line. The set-oriented geometric statement that corresponds to this physical model is the following postulate.

POSTULATE 3.1

On two distinct points in space there lies exactly one line.

This postulate is often stated as the familiar: "Two points determine a line."

Even though the mathematics to this point has been relatively informal, this informality does not allow the luxury of being imprecise in our language. After all, it is this exactness in language that gives mathematics its universal appeal. The word "determines" should be clearly understood. It does not simply mean "gives us" or "insures the existence of." As we saw above, on any given point there exist infinitely many lines, and so a broad interpretation of "determines" would allow us to say that a point determines a line. Rather, we demand that it convey the meaning of uniqueness, distinctness, one and only one, etc. When we say that two points determine a line, we mean they determine a unique line and no other. Thus, a single point cannot determine a line in the sense that we define "determine."

DEFINITION 3.2

The term **collinear** is given to all points that lie on a given line.

In other words, if l is a line and S is a set of points, and if $S \subseteq l$, then the points of S are collinear.

An immediate consequence of the fact that two points determine a line is the following, which may be classified as a theorem.

THEOREM 3.2

On two distinct intersecting lines in space there lies exactly one point.

The agreement on the use of the word "on" allowed us to state Theorem 3.2 in language almost identical to Postulate 3.1. In addition to the interchange of the words "line" and "point" in the two statements, there is also the inclusion of the word "intersecting" in the second. This is a major difference. In the first st ment, there are no restrictions on the two distinct points; they may li. arywhere in space and they will still determine a unique line. Not so with the two lines; they must intersect in order to determine a unique point.

The truth of Theorem 3.2 can be demonstrated quite readily. Suppose two distinct lines, l_1 and l_2, intersect in two distinct points, P and Q. This fact would be in direct contradiction to the first statement, which assures us that exactly one line lies on the two points P and Q. Thus l_1 and l_2 cannot intersect in more than one point.

Consider now the situation where the two lines do not intersect. There are two separate cases to examine. The first is the case when the two nonintersecting lines lie in the same plane. This is the familiar case of parallel lines. Parallel lines may be defined in a formal way as follows:

DEFINITION 3.3

Two lines, m_1 and m_2, are said to be **parallel** if and only if they lie in the same plane and $m_1 \cap m_2 = \varnothing$.

The second case occurs when the two lines lie in different planes.

DEFINITION 3.4

Two lines, m_1 and m_2, are said to be **skew** lines if and only if they lie in different planes and $m_1 \cap m_2 = \varnothing$.

A possible physical model of this situation is to think of the line determined by the 50-yard line of a football field and another line determined by the vertical support of the goal post at one end of the field.

3.2   LINES AND PLANES

We shall now extend our discussion of intersections of sets of points by considering the two-dimensional set of points, namely the plane. Recall that a plane is a particular set of points with the property of "flatness." Recall also the fact mentioned in Section 2.5 that if points A and B lie in a plane, then $\overleftrightarrow{AB}$ lies in the plane. Since a plane is a set of points, we can quickly dispose of the intersection of a set consisting of a single point with a plane by noting that the intersection is either the empty set or the set consisting of the single point. Of course, there are infinitely many planes lying on a given point as well as infinitely many points on a given plane.

How about the intersection of a line with a given plane? There are two main cases to be considered here. Either the intersection is the empty set or it is not the empty set. If the intersection is the empty set, then we shall say that the line is parallel to the plane. A physical interpretation of this would be to say that a line lying in the plane determined by the floor of a room is parallel to the plane determined by the ceiling of that room (provided it is a "box-shaped" room and not one in a modernistic house!).

If the intersection of a line with a plane is not the empty set, then there are two subcases; they need only be mentioned since they have already been considered. The first case is when the intersection is but a single point—when the line "pierces through" the plane as pictured in Fig. 3-1(a). The second case happens when the intersection is more than one point. Then, as we have noted, the intersection is the line itself. This is shown in Fig. 3-1(b).

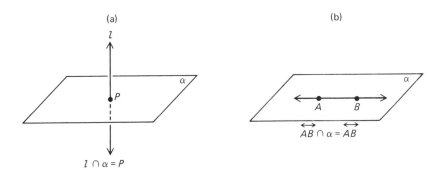

Figure 3-1

What does it take to "determine" a plane? In case you do not already know the answer, picture a familiar spiral notebook. Imagine passing a long pencil or a thin rod down through the spiral that protrudes from the ends of the pages and covers of the notebook. Holding the ends of the pencil or the rod in your two hands, you could cause the pages and/or the entire notebook to spin around the pencil as an axis, so that the pages or the notebook would rotate a full circle or more. Thinking of the pencil or the rod as a physical model of a line and a page of the notebook as a model of a plane, we may conclude that a line is not enough to determine a unique plane. In fact, there are infinitely many planes lying on a given line.

Returning to the rotating notebook, if you can further imagine a friend coming up and placing the tip of her index finger in the path of the rotating book, then you have conjured up a physical answer to the question posed in the first line of the preceding paragraph. With the finger in the way, the plane of the notebook is no longer free to occupy any of its infinitely many possible positions; rather, it becomes fixed in the same way a second point fixed the rotating line of the previous section. This is a plausible justification of the following postulate.

POSTULATE 3.3

Given a line and a point not on the line, there is exactly one plane that lies on the line and on the point.

EXERCISE SET 3.1

1. Explain why in the introductory paragraphs of this chapter the notation $A \in l$ was used when reference was made to a point lying on a line while the notation $\overleftrightarrow{AB} \subseteq \alpha$ was used with reference to a line lying on a plane.

2. The word "plane" was taken as an undefined term. Would it have been possible to define the word "plane" in terms of the word "line?" If your answer is "yes," try to formulate such a definition.

3. Give two other physical descriptions of skew lines different from the one used in the text.

4. Give a plausible argument why the following statement is true: Three noncollinear points determine exactly one plane.

5. Do the same for: Two intersecting lines determine exactly one plane.

6. Do the same for: Two parallel lines determine exactly one plane.

7. Can two skew lines determine a plane?

8. Think of three distinct lines in a plane. Draw pictures representing these lines where there are no points of intersection; one point of intersection; two points of intersection; three points of intersection.

9. The statement in Exercise 4 above is sometimes referred to as the "three-legged stool" property. Do you see why? Explain why it is not

necessary for a four-legged chair to "sit" exactly balanced on a level floor. If you had a four-legged chair that did not "sit" properly on a level floor and you decided to remedy the situation by sawing off a portion of one of the legs, how would you go about deciding exactly where to make your cut with the saw?

3.3 PLANES AND PLANES

In this section, we shall consider the possible relationships of two or more distinct planes. We are getting away from the two-dimensional object, which is the plane, and into three dimensions. Many call this **solid** geometry as opposed to **plane** geometry. In this section, only the slightest introduction to solid geometry will be given. It will be given a fuller treatment in a later chapter.

When we consider the intersection of two planes, there are but two possibilities to consider. Either the intersection is vacuous or it is not. If the intersection is the empty set, then the two planes are said to be parallel. This concept may be extended to more than two planes. For example, if three planes, α, β, and γ, are pairwise disjoint— that is, $\alpha \cap \beta = \varnothing$, $\alpha \cap \gamma = \varnothing$, and $\beta \cap \gamma = \varnothing$ —then α, β, and γ are said to be parallel.

On the other hand, if two planes have a nonempty intersection, then their intersection is a line. Try to think of any other possible situation if you are not convinced that this is the case. If three planes, α, β, and γ, are *not* pairwise disjoint, then various possibilities may occur. Some of these are left to the exercises. However, you need look no further than the edges and corners of the room in which you are sitting in order to visualize some of those possibilities.

EXERCISE SET 3.2

1. Take a piece of notebook paper and fold it lengthwise so that the vertical crease of the fold is parallel to the edges of the paper. The two "half-planes" of the paper should be in such a position that they define two distinct planes. Consider the flat surface of a desk as a representative of a third plane. The paper may be placed on the desk in three different positions so that no two of the three "planes" coincide. Describe these by an English sentence. Then, by giving appropriate names to the "planes" and the "lines" of intersection, describe these positions by means of set notation.

2. If three distinct points, A, B, and C, are such that the three lie in each of two distinct planes π and ρ, what can be said about the points A, B, and C?

3. If W, X, Y, and Z are four distinct points not all on the same plane and if no three of them are collinear, how many distinct planes are determined by these four points?

4. Be prepared to discuss whether or not it makes sense to talk about "determining" space in the same sense as it does to "determine" a point, a line, and a plane. If it does seem to make sense, be ready to tell how it could be done.

3.4 ANGLES

The idea of an angle is familiar to most persons and a picture of one could probably be drawn if requested. It shall be defined here in set language, which will be free from the idea of measurement.

DEFINITION 3.5

An **angle** is a set of points that is formed by the union of two rays having a common endpoint.

An immediate consequence of the definition is that an angle is a subset of a plane. Why? A picture of an angle formed by the rays $\overrightarrow{BA}$ and $\overrightarrow{BC}$ is given in Fig. 3-2. The common endpoint is called the **vertex** of the angle

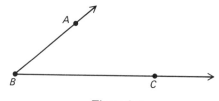

Figure 3-2

and each ray is called a **side** of the angle. The notation for angle is $\angle$; one writes $\angle ABC$ with the agreement that the letter naming the vertex be placed between the other two letters that name points lying on each of the two sides of the angle. Thus, we may properly write $\angle ABC = \angle CBA$.

It is important to realize that the definition of angle makes no restriction on the relative position of the two rays. It is possible for rays $\overrightarrow{BA}$ and $\overrightarrow{BC}$ to be opposite rays, as shown in Fig. 3-3(a). In this case, the angle is called a **straight angle**. It is also possible for $\overrightarrow{BA}$ and $\overrightarrow{BC}$ to be the same ray as shown in Fig. 3-3(b). Such an angle will be called the **zero angle**. However, given a set of points such as those pictured in Fig. 3-3(a), it

(a)   (b)

Straight angle $\angle ABC$ Zero angle $\angle ABC$

Figure 3-3

would be difficult to decide whether the set of points was simply a line or an angle. Therefore, it is the usual convention in the study of incidence geometry to exclude consideration of both the straight angle and the zero angle although these angles do have a place in the study of geometry and trigonometry when the notion of measurement is considered. In this text, the word "angle" will be used to denote an angle that is neither a straight angle nor a zero angle unless noted otherwise.

An angle has the property of separating the plane in which it lies into three distinct and mutually disjoint subsets. One of the sets is the angle itself, and the other two are called the **interior** of the angle and the **exterior** of the angle. In Fig. 3-4, the interior of $\angle MPQ$ is represented by the shaded

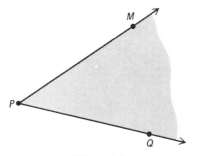

Figure 3-4

region and the exterior is represented by the unshaded region. The interior may be defined in a precise yet simple manner. Rather than give it here, you will be asked to formulate the definition in one of the exercises that follow.

Special mention should be made of several instances where angles occur in pairs. The first is the case of angles formed by two intersecting lines, as shown in Fig. 3-5. The point of intersection of the two lines is

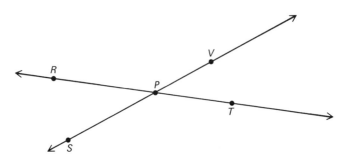

Figure 3-5

the vertex of the angles, and two distinct pairs of angles are formed, which are called **vertical angles**. In Fig. 3-5, ∠RPS and ∠VPT are one pair of vertical angles and ∠RPV and ∠SPT are the other. The sides of the angles are rays whose endpoints are the point P. Notice that the angles that make up the pair of vertical angles do not have a ray in common.

Adjacent angles are a pair of angles that do share a common side. They are such that they lie in the same plane (**coplanar** is the technical term) and have a common vertex and a common side that must be "between" the two angles. A picture of this situation is shown in Fig. 3-6. In this

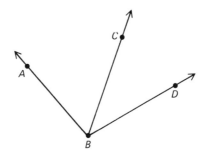

Figure 3-6

figure there are shown three rays sharing the same endpoint B. The three rays, $\overrightarrow{BA}$, $\overrightarrow{BC}$, and $\overrightarrow{BD}$, actually form three distinct angles, ∠ABC, ∠ABD, and ∠CBD, and therefore three distinct pairs of angles, ∠ABC and ∠ABD, ∠ABC and ∠CBD, and ∠CBD and ∠ABD. Each of these pairs shares a common ray; for the first named pair it is $\overrightarrow{BA}$, for the second it is

$\overrightarrow{BC}$, and for the third it is $\overrightarrow{BD}$. Yet, only in the second case is the common

ray $\overrightarrow{BC}$ "between" the two angles $\angle ABC$ and $\angle CBD$. Thus, only this pair may be called adjacent angles. The word "between" has been put in quotation marks to suggest that the idea of being between two angles must be for the moment intuitive, an idea that makes sense to the reader but that has not been defined in a proper fashion. This definition can be made using the idea of the interior of an angle but will be left to the exercises.

A word of warning to the reader! Do not confuse the set of points that is an angle with the set of points that is the interior of the angle. These two sets are disjoint. For example, in Fig. 3-6 you should not think that since $\angle ABC$ and $\angle CBD$ are in some sense "smaller" or "inside" $\angle ABD$, each is a subset of $\angle ABD$. Neither $\angle ABC$ nor $\angle CBD$ is a subset of $\angle ABD$!

Finally, if two adjacent angles are such that their noncommon sides are opposite rays, then these angles are said to form a **linear pair.** Another way of stating this relationship is to say if two adjacent angles, $\angle ABC$ and $\angle CBD$, are such that A, B, and D are collinear, then the two angles form a linear pair. This relationship is pictured in Fig. 3-7.

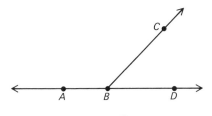

Figure 3-7

In some contexts such angles are called supplementary angles. However, the term "supplementary" usually carries a connotation of size or measurement. As such, it has no place in a discussion of incidence geometry and so will not be used in this connection. Moreover, supplementary angles need not be adjacent, whereas a linear pair must be.

EXERCISE SET 3.3

1. In the figure on page 44 are four distinct concurrent rays. How many distinct angles are pictured? Name them. How many distinct pairs of adjacent angles are pictured? Name them.

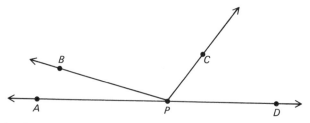

2. If, in the figure shown in Exercise 1, the points A, P, and D are collinear, how many linear pairs are pictured? Name them.

3. Consider a situation with five concurrent rays in a plane. Without actually drawing a picture, answer the following: How many distinct angles are determined? How many distinct pairs of adjacent angles will there be? Does the answer depend on whether all rays are either in a half-plane or a subset of the boundary of that half-plane?

4. Describe the interior of an angle by making use of the half-planes formed by the lines of which the sides of the angle are subsets.

5. Describe adjacent angles by making use of their interiors.

6. Describe the interior of a straight angle; of a zero angle.

7. Does your answer to the above exercise suggest another reason why a straight angle and a zero angle are not usually considered in a study of incidence geometry?

8. Go back and take another look at Fig. 3-4. Suppose that the shaded region of the figure had been called the exterior of the angle and the unshaded portion the interior. Would such an agreement cause any difficulties that would make your answers to Exercises 4 and 5 any different?

SUGGESTED READINGS

[6], Ch. 1

[13], Ch. 2

[21], Ch. 2

[25], Ch. 1

[26], Ch. 2

4

MEASUREMENT OF SEGMENTS AND ANGLES

In the present chapter, we shall begin an examination of the so-called **metric** properties of sets of points. Recall that the previous chapters have dealt with nonmetric geometry and some of the ideas connected with it. We shall see that some of these ideas may be restated in terms of metric ideas. As the chapter title suggests, the main topic of interest is the measurement of segments and angles that can best be described as a one-dimensional operation. Two- and three-dimensional measurement as well as the measurement associated with circles will be dealt with in subsequent chapters.

4.1 WHAT IS MEASUREMENT?

Measurement is a most important topic in mathematics, because it is a process that unites in a very close manner the concepts of geometry with those of arithmetic. Basically, measurement assigns a positive number to a set of points. This number is called the **measure** of that set. The study of measurement tells how this assignment takes place. As you might expect, since it does unite geometry and arithmetic, measurement theory has become a subject of great interest to mathematicians. Our exposition could become quite technical and highly theoretical. However, in keeping with the spirit of this text, the discussion of measurement will be rather informal and will rely on familiar experiences in measuring real objects.

Perhaps it might be instructive to create a hypothetical example of the use of measurement of real objects. Throughout this example and the rest of the chapter as well, it will be assumed that to everyone the word *length* means a particular property that belongs to line segments.

Suppose that two recent graduates are about to start their first jobs. The two, Lydia Longfoot and Pamela Petite, have found a suitable apartment in which to live. They are in the process of moving in and are making decisions regarding where to place the various articles of furniture that they have brought along. Before moving the heavy furniture all about, the two wisely decide to measure the articles of furniture and the intended locations of placement to see if they will *fit*. Not having a ruler or yardstick handy, Lydia decides it will be just as easy to step off the distance. So, in a manner familiar to most of us, she places the heel of one foot at the beginning point and then, placing the heel of her other foot at the toe of her first, and so on, she steps off the length of the sofa and proclaims its length to be "a little more than eight feet long." As she is about to measure the place where the sofa is to be put, the telephone rings, and it is for her. Being impatient to get settled, Pamela decides to continue the job, and so she steps off the length of the place where the sofa is to be put in a similar measuring manner. She finds the distance to be "just about nine feet long." So, feeling that eight feet is less than nine feet, they go ahead and start to move the heavy sofa into its place. They are surprised to find that the sofa does not fit as expected. Being duly chagrined, they sit down to try to decide where they went wrong and soon come up with the answer. Sadder but wiser, they place the sofa in another position.

Next, they set about to see if the desk will fit through the doorway into the bedroom. Not wanting to damage the woodwork by jamming the desk up against it, they again decide that prior measurment is the best idea. Abandoning the "foot" measurement, Pamela comes up with the bright idea to use a piece of string that happens to be handy. Stretching the string tightly and placing her fingers on the string at the points where the opposite edges of the desk touched the string, she turns to the doorway and measures it by using the stretched string. Since the length of the string is less than the width of the doorway, she proudly proclaims that the desk will go through without harming the woodwork. And so it does!

Finally, they feel that some new curtains will make their apartment seem more cheerful. Having had experience in these matters before, and, besides, not wanting to appear foolish before a salesclerk, they properly decide that the best course of action is to wait until they can use a standardized measuring instrument, such as a yardstick or 12-inch ruler, to measure their windows as accurately as they can and then go to the store with these measurements.

Maybe the above story is not too fictitious. At any rate, it points up some of the basic ideas connected with measurement of real objects. In assigning a number to the length of the sofa, Lydia decided on the number by means of comparing the length of the sofa to the length of her foot. She did this by placing her feet *end-to-end* and finding the number of times the length of her foot was contained in the length of the sofa. We say that her foot was used as the **unit of measure**. Even though she did not actually do it, we can properly surmise that if Pamela had measured the length of the sofa by using her foot as the unit of measure, a different number would have been assigned as the measure of the sofa. Which measure was more correct? If we disregard for the moment their lack of precision as indicated by the phrases "a little more than" and "just about," both are correct. The girls' mistake was in comparing the two measures when *different units of measure* were used.

The second example of the desk and the doorway was given simply to point out that other units of measure can be used. In this case, it was the segment exemplified by the stretched string marked off between the two points by her fingers.

The last example of the curtains gives the lesson that most commonly accepted units of measure are the standardized ones that are agreed upon by all persons. Since mathematics is a means of communication, it is essential that we speak the same language. In the specific realm of measurement where we wish to communicate and to make comparisons, it is necessary that standardized units be used. There are two systems of standardized measurement used throughout the world. In the United States the **English** system of measurement is used. Nearly all other countries use the **metric** system. A special section of this chapter is devoted to the metric system.

The other major lesson to be learned from the above tale is that measurement of real objects is never exact but must be an approximation. Perhaps the two feet of Lydia or those of Pamela are not of the same length. The physical act of measurement usually involves the placement of the measuring device alongside or on top of the object that is to be measured. Since a measuring instrument is usually a manufactured item, it is subject to error. Secondly, the act of measurement involves a reading of the measuring instrument. This reading is also subject to error. It is true that modern instruments of measurement have become very refined and that the amount of error can be made very small, but the fact remains that some error in measurement is present. This lack of exactness does not cause any difficulty in actual practice, however, because man has learned how to adjust to this fact of life. A more complete discussion of precision and error in measurement is given in a later portion of this chapter.

Our theory of measurement of the abstract sets of points that are the subject matter of this course will be based upon actual measurement of real objects. One major difference occurs however. Since the abstractions exist only in our minds, we may become highly theoretical if we wish. One immediate consequence is that measurement of geometric objects becomes exact, and there is no error involved. A reflection upon the real world will tell us that this makes sense. Simply because our feeble measuring devices are unable to measure an object exactly does not mean that that object does not have an exact measure. It does, but we are unable to find out what that measure is; we merely make as good an approximation as is possible. All we are saying is that in the abstract mathematical sense we *are* able to find out what the exact measure is. In fact, it turns out in many instances that we are not interested in the exact numerical measure of an object, but rather that its measure is exactly equal to that of another object of the same type. For example, in the well-known Isosceles Triangle Theorem, the interest lies not in the numerical measure of the two sides and the two angles, but in the fact that these measures are exactly the same for the two sides and for the two angles.

Thus, it may be that the numerical measure of an object may be any real number; for example, a given segment $\overline{AB}$ may be $\sqrt{2}$ in. long or another may be $(6 + \pi)$ ft. long. What is needed to find these measures of our abstract objects is an abstract mathematical measuring device; for segments a "mathematical ruler" would be needed. It turns out that the familiar real number line serves the purpose very well.

EXERCISE SET 4.1

1. Take a half sheet of ordinary notebook paper and cut or tear it lengthwise into five strips of paper. Use a straightedge (not a ruler!) and a pencil to mark on the first strip a line segment that you estimate to be one inch long. Turn the strip over so you cannot see what you have drawn. Repeat the process on the other four pieces of paper insuring that you do not see what you have done on the previous ones. Now,

using a ruler marked in inches, measure each of your line segments to see how close you came.

2. Use the other half of the piece of notebook paper left over from Exercise 1 to repeat the experiment, this time trying to draw a segment three inches in length.

3. Use a fairly large piece of paper (an old newspaper might do) and mark off the front and back of your bare right foot. Then use a ruler to compare your "foot" to the standarized "foot" by measuring the length of your foot as marked on the paper. Repeat this twice more to see if you get the same measurement each time. Now try it with your left foot.

4. Which length—inch, foot, yard, mile, or other—would be most suitable to use as a unit of measure if you were measuring:

(a) the width of a desk
(b) the height of a room
(c) the distance across a playground
(d) the size of notebook paper
(e) the distance to the moon
(f) the thickness of a sheet of glass

(g) the height of a three-year-old child
(h) the height of a twenty-year-old girl
(i) the length of a mosquito's leg
(j) the size of your foot

4.2 THE NUMBER LINE AND THE REAL NUMBERS

The **number line** may be thought of as a pictorial device to enable a person to visualize some of the relationships that exist between real numbers. Figure 4-1 shows several ways in which the number line can be pictured. Part (b) of Figure 4-1 shows a way that is not typically used simply because we happen to live in a right-handed world. It is not incorrect however. Part (c) shows a number line pictured by someone who "thinks big." It is not wrong either. In fact, all of them are correct representations of the number line. Since there is only one set of real numbers, we should agree that there is only one number line, even though there may be many different ways of drawing it. Looking at parts (a) and (c) of Fig. 4-1, we notice the difference in the lengths of the segments between the points marked by 0 and 1. This does not mean that there are more points in the segment of part (c) than there are in the corresponding segment in part (a). There is an infinite number of points in each. Moreover, the measure of length (or the measure of anything else, for that matter) has nothing to do with the

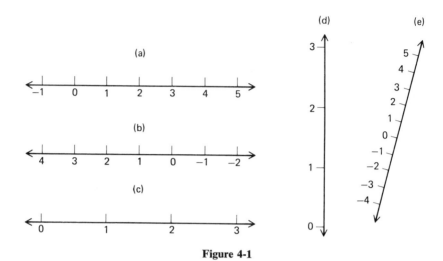

Figure 4-1

number of points in a set.   It is impossible to count the number of points in a segment.

What then is the number line?   It is best thought of as an association between the points of some arbitrary line and the set of real numbers.   This association is no more than a one-to-one correspondence between the points and the numbers such that to each point there corresponds exactly one **real** number and to each real number there corresponds exactly one point.

It might be well at this stage to briefly review the set of **real numbers** and those properties of it that have particular interest in our study of measurement and geometry.   Recall that the reals have some especially noteworthy subsets.   The most basic subset is that of the **natural numbers** which is also known by the name of the **positive integers.**   It is the set $N = \{1, 2, 3, 4, \ldots\}$.   Another subset is the set of **integers,** which is

$$I = \{\ldots\ -4,\ -3,\ -2,\ -1, 0, 1, 2, 3, 4\ldots\}$$

The integers are often subdivided into three distinct and disjoint subsets, the negative integers, $\{\,0\,\}$, and the positive integers.   Still another is the set of **rational numbers**, which is the set of all numbers of the form a/b, where a and b are integers and $b \neq 0$.   Notice that those rationals where $b = 1$ form a set of numbers that behave exactly like the set of integers.   Thus, while it is not precisely correct from a purely theoretical point of view to say that the integers are a subset of the rationals (and thus of the reals), it is essentially correct when speaking from a practical or operational point of view.[1] Finally, there is the set of the **irrational numbers,** which are simply defined as those real numbers that are not rational. The classic proof that $\sqrt{2}$ is irrational is probably familiar to most students and will not be repeated here. A meaningful interpretation is that which associates the irrationals with the set of nonrepeating, nonterminating decimals and the rationals with the set of repeating and/or terminating decimals. In summary, then, the real numbers are the union of two disjoint sets, the rationals and the irrationals. The rationals have two subsets of note, the integers and the natural numbers.

Turning now to some of the properties of the reals, we begin by stating that there are two familiar binary operations defined on the reals, addition and multiplication, which are symbolized in the usual way. Belonging to these operations are certain fundamental properties. The letters a, b, and c stand for any real number, not necessarily distinct.

I. **Closure:**

$$a + b \text{ is a real number}$$

$$ab \text{ is a real number}$$

II. **Commutativity:**

$$a + b = b + a$$

$$ab = ba$$

III. **Associativity:**

$$(a + b) + c = a + (b + c)$$

$$(ab)c = a(bc)$$

IV. Existence of **Identity Elements:**

There is a real number 0 such that $a + 0 = 0 + a = a$

There is a real number 1 such that $a \cdot 1 = 1 \cdot a = a$

---

[1] This same point of view is true about the natural numbers and the positive integers. Theoretically, they are not the same set, but since they act in exactly the same way and have exacly the same properties, most persons treat them as the same set.

V. Existence of **Inverse Elements:**

For all a there exists a number $-a$, called the **additive inverse** of a, such that

$$a + (-a) = (-a) + a = 0$$

For all nonzero a there exists a number a^{-1}, called the **multiplicative inverse** of a, such that

$$a \cdot a^{-1} = a^{-1} \cdot a = 1$$

VI. The **Distributive Property:**

For all a, b, and c:

$$a(b + c) = ab + ac$$

Notice that for the first five properties, the structure for each of the operations is the same, with exception of the zero in multiplication having no inverse. This structure is the structure that belongs to what is known as a **commutative group.** When all eleven properties hold as they do here for the set of real numbers and the two operations, the system is known as a **field.**

In view of the discussion of the previous paragraphs, perhaps a better pictorial representation of the number line may be given. Figure 4-2 is such

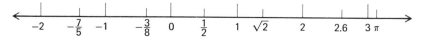

Figure 4-2

an attempt. Here are shown some of the infinitely many numbers that are not integers. If an unwary reader were only shown Fig. 4-1, he might be led to think that the number line has only integral points. Such is certainly not the case!

The "$=$" sign used between two expressions using real number symbols means that they are expressions for the same number. This relation of equality between two real number symbols is another example of an equivalence relation and thus has the properties of being reflexive, symmetric, and transitive. There are other properties of equality that could be classified as theorems; but they shall be simply stated here with the assumption that they are either familiar or, if not, that they will be intuitively plausible statements.

I. Addition Property of Equality:

For all a, b, and c: If $a = b$, then $a + c = b + c$.

II. Multiplication Property of Equality:

For all a, b, and c: If $a = b$, then $ac = bc$.

III. Cancellation Properties:

For all a, b, and c: If $a + c = b + c$, then $a = b$.

For all a, b, and $c \neq 0$: If $ac = bc$, then $a = b$.

Perhaps of more interest are the two relations of inequality, "is greater than" and "is less than." The properties belonging to these relations are called the **order properties** for real numbers. On the assumption that these relations and properties may not be as familiar as those of equality, more space will be devoted to their review.

DEFINITION 4.1

For any two real numbers a and b, a **is greater than** b (written: $a > b$) if and only if $a - b$ is a positive number, where "$-$" stands for ordinary subtraction.

EXAMPLES

(a) $8 > 5$ because $8 - 5 = 3$, which is a positive number.

(b) $3 > -2$ because $3 - (-2) = 5$, which is a positive number.

(c) $-1 > -7$ because $-1 - (-7) = 6$, which is a positive number.

Putting this relationship in geometric terms, we refer to the number line of Fig. 4-2, where the positive direction is to the right, and state that one number is greater than a second if the first is to the right of the second.

DEFINITION 4.2

For any two real numbers a and b, a **is less than** b (written: $a < b$) if and only if b is greater than a.

This second definition states that there is a very close relationship between the two relations of "is greater than" and "is less than" and that we need to examine but one of them in any detail. For if two numbers m and n are unequal, then one must be greater than the other; that is, even if we write $m > n$ and read it "m is greater than n," we can also interpret it to say "n is less than m."

The first order property is stated below as a postulate.

POSTULATE 4.1   (The **Trichotomy** Postulate)

Given two real numbers, a and b, one and only one of the following is true:

$$(1) \quad a = b; \quad\quad (2) \quad a > b; \quad\quad (3) \quad a < b$$

Do you see why the word "trichotomy" is a good one to describe this property?

There are many other properties dealing with inequalities. Only two of them are given here; some of the others are left to the exercises.

I. **Addition Property of $>$:**

For all a, b, and c: If $a > b$, then $a + c > b + c$.

II. **Multiplication Property of $>$:**

For all a, b: If $a > b$, then $ac > bc$, provided $c > 0$.

For all a, b: If $a > b$, then $ac < bc$, provided $c < 0$.

Each of these properties needs some amplification, although it is probably true that the addition property is much easier to visualize. Suppose that $a > b$. Then a would be to the right of b on the number line. This is shown in Fig. 4-3(a). Imagine a small two-pronged pointer arranged as pictured, so that it points out the relative positions of a and b. If c is a positive number, then a, b, and the little pointer are all displaced to the right, as shown in Fig. 4-3(b). If c is a negative number, the displacement is to the left, as shown in Fig. 4-3(c). If c is 0, there is no displacement. In all three cases, the relative positions of $a + c$ and $b + c$ are the same as the original a and b.

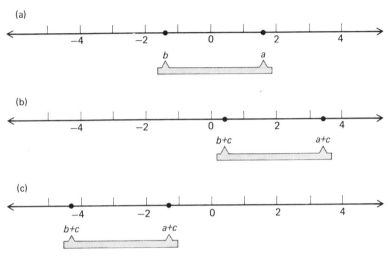

Figure 4-3

The multiplication properties are more easily understood by means of numerical examples. Begin with the true statement: $9 > 4$. Multiply the numbers on each side by the positive number 3. The resulting numbers are 27 and 12, and the order relation between these two is the same as the original relation between 9 and 4; that is, $27 > 12$. Now take the same inequality $9 > 4$ and multiply the two numbers by the negative number -2. The resulting numbers, -18 and -8, have an order relationship that is the reverse of the original; that is, $-18 < -8$. If you do not feel convinced you should work more examples for yourself until you are convinced of the truth of the multiplication properties. Notice that neither of the two statements allows the multiplier c to be equal to 0. What happens if c does equal 0?

Many times it is useful to combine two statements of an order relationship into a single compound statement. The most common one is to combine the statement $a > b$ with the statement $b > c$ to produce the compound statement using the connective "and" and say "$a > b$ and $b > c$," which is written $a > b > c$. It seems entirely proper to say that if this joint order relationship does hold, then "the real number b is between a and c."

DEFINITION 4.3

A real number b is said to be **between** the real numbers a and c if and only if either $a > b > c$ or $a < b < c$ holds.

A very important fact to take note of at this point is that if we are given any two distinct real numbers a and b, there is always a third real number that is between those two, no matter how close together the two original numbers were chosen. One can always choose the average of the two numbers as the one that is between them. This property of the real numbers—between any two there is always another—is called the property of **density**.

Another common combination of two order statements is the compound statement using the connective "or" with the statements "$a > b$" and "$a = b$." This compound statement is symbolized by $a \geq b$ and is read "a is greater than or equal to b." Recall the mathematical interpretation of the word "or" that is the inclusive meaning. For the statement $a \geq b$ to be true, either one or the other or both of the simple statements must be true. Since the trichotomy postulate precludes the possibility of both being true, we then agree that $a \geq b$ will be true if $a > b$ or $a = b$. In a similar fashion, $a \leq b$ means "a is less than or equal to b."

EXERCISE SET 4.2

1. Given the set $A = \{0, 1, 2\}$.
   Is this set closed with respect to the operation of ordinary addition? Ordinary multiplication?

2. Let B be the set of normal activities of people and let the operation on the elements of this set be "is followed by." Given pairs of elements of this set, the operation is not commutative if the final net result of the two activities is different. For example: putting on socks followed by putting on shoes is not the same as putting on shoes followed by putting on socks. Another example: opening a window followed by throwing a large stone through the space is not commutative. Give five more pairs of activities that do not commute with respect to this operation.

3. Tell which property the following examples illustrate:
   (a) $(4 + 2) + 13 = 4 + (2 + 13)$
   (b) $\frac{3}{8} \cdot \frac{8}{3} = 1$
   (c) $2 \cdot \frac{1}{2} - 5 \cdot \frac{1}{2} = (2 - 5) \cdot \frac{1}{2}$

(d) $-\frac{5}{7} + [-(-\frac{5}{7})] = 0$

(e) $(9 \cdot 0.6) \cdot \frac{4}{13} = 9 \cdot (0.6 \cdot \frac{4}{13})$

(f) $11(9 - 4) = 11 \cdot 9 - 11 \cdot 4$

(g) $(8 + 2) + (6 + 1) = 8 + [2 + (6 + 1)]$

(h) $(8 + 2) + (6 + 1) = (6 + 1) + (8 + 2)$

(i) $(5 - \pi) \cdot 1 = (5 - \pi)$

(j) $0 + 3\frac{1}{5} = 3\frac{1}{5}$

4. Is the relation "is greater than" an equivalence relation?   If so, try to prove this fact.   If not, tell which of the three properties hold and which do not.

5. Use the order properties for real numbers to "solve" the following:

(a) $x + 6 > 9$

(b) $x - 11 < 23$

(c) $3x \geq 18$

(d) $\dfrac{x}{6} > 1.5$

(e) $5x - 9 \leq 4$

(f) $\frac{1}{2}x + 2 < 3$

(g) $3 - 5x \geq 17$

(h) $\dfrac{1 - x}{4} < 2$

(i) $3 > 2x + 1 \geq -2$

(j) $-4 < \dfrac{x - 6}{2} < 7$

(k) $13 \geq 5 - 6x \geq 1$

6. Make up five specific numerical examples of the relation $a > b$.   Try to vary these, so that not all numbers of the five pairs are positive. Examine the order relation of the additive inverses of these pairs. Make a conjecture regarding the results.   Try to prove this conjecture or at least give a reasonable argument explaining why the conjecture is true.

7. Repeat Exercise 6 with five new pairs of numbers and their multiplicative inverses.

8. Explain why the following statement is true: "If $a > b$ and $c > d$, then $a + c > b + d$."

9. Does the set of integers have the property of density? Explain your answer.

10. What does the statement $a > b > c$ tell us about the order relation of a and c?   Why?

11. It is possible (although it is rarely, if ever, done!) to combine the statements $a > b$ and $a > c$ into the compound statement $c < a > b$, which would read "a is greater than b and a is greater than c." Does such a compound statement tell us anything about the order relation of b and c?   Is it correct to interpret the statement $c < a > b$ as saying "a is between b and c"?

12. If $a > b$, try to prove that the average of a and b is between a and b by using the addition and multiplication properties of $>$. (*Hint:* To show $a > \dfrac{a+b}{2} > b$, there are two inequalities to verify: $a > \dfrac{a+b}{2}$ and $\dfrac{a+b}{2} > b$. Prove each one separately. Begin the first one by taking the inequality $a > b$, and then add the number a to both sides; the second is started the same way, except add the number b to both sides.)

13. We have seen that there exists at least one number between two distinct real numbers. How many numbers are there between any two distinct real numbers?

4.3   MEASURING A SEGMENT

As suggested earlier, we wish to develop a basis for assigning a positive number to a set of points that is a segment. The segment is an abstraction that exists in our mind, even though it is an excellent mathematical model for many real objects that have a straight edge for a side. We wish the mathematical measuring system to be consistent with our methods for measuring real objects. The mathematical ruler for measuring segments will be the number line. This section tells how to use this ruler.

Two main problems confront us. First, how do we place the ruler alongside the segment so as to be able to read the measure? Secondly, after the ruler has been properly placed, how do we read the measure? Before we attempt to answer these questions, there is a matter of terminology and mathematical notation that should be clarified. Notice that the word "measure" rather than "length" is being used in relation to segments. The measure of a segment does indeed give us the length of a segment, because length is a property of segments. However, when we turn our attention to angles, the concept of length has no meaning, but the word "measure" does make sense. In other words, "measure" is a more general term and can be used in both contexts; that is why the word "measure" is preferred. When we refer to the measure of a segment, the notation $m(\overline{AB})$ will be used; for an angle, it will be $m(\angle ABC)$.

Consider some given segment $\overline{AB}$, where A is distinct from B. These two endpoints also determine the line $\overleftrightarrow{AB}$. We might just as well let this line be our number line, since we agreed that the number line could be any arbitrary line that has its points in a one-to-one correspondence with the real numbers. Consequently, the point A corresponds to some real number, say, a, and B corresponds to some number b. The numbers a and b are called the **coordinates** of A and B, respectively. One of the two directions of the number line must be the positive one, and since A and B are distinct points, one of them must be further in the positive direction than the other. For the moment, let us suppose it is B. Then the measure of the segment

$\overline{AB}$ is $m(\overline{AB}) = b - a$. This relationship is pictured in Fig. 4-4. We are assured that $b - a$ is a positive number, because $b > a$, since b is further in the positive direction than a.

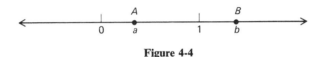

Figure 4-4

This tells us how to place the mathematical ruler, but it in no way tells

how to read the ruler. In other words, we can find $m(\overline{AB})$ if we know precisely what the numbers a and b are. If a and b are located on the number line, as shown in Fig. 4-4, then they are both positive. However, Fig. 4-5(a) shows a placement where a is negative and b is positive, whereas Fig. 4-5(b) shows a different placement where both are negative.

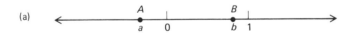

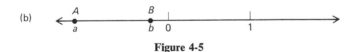

Figure 4-5

Thus, the values of a and b depend on the placement of A and B relative to the point that is labeled with the number 0. (This point is called the **origin**.) A logical question to ask is whether these different placements have any effect upon the value of $b - a$. The answer will be postponed for the moment.

Instead, you will be asked to ponder the following problem. Examine Fig. 4-6(a), which is the same as Fig. 4-4, and then look at Fig. 4-6(b). In the second figure, A and B are placed in exactly the same position relative

(a)

(b)

Figure 4-6

to the origin, but the point labeled with the number "1" is to the right of B rather than to the left. The problem to be resolved is whether this placement has any effect upon the value of $b - a$.

Let us answer this question and the prior question at the same time. Even though we are now in the realm of a theoretical, abstract system, we can answer these questions by relating back to our ideas of measurement in the real world. In all four of the preceding diagrams, the length of $\overline{AB}$ is the same. In Fig. 4-5(a), Fig. 4-5(b), and Fig. 4-6(a), the measure of $\overline{AB}$, that is, $b - a$, is the same. The analogy in real measurement is this: suppose, for purposes of explanation, $\overline{AB}$ is 1 in. in length. Then the three figures would be similar to placing an inch-ruler alongside $\overline{AB}$ with A at the point marked 1 in., in which case B would lie at the point marked 2 in., or placing A at the mark for $2\frac{1}{2}$ in., in which case B would lie at the mark for $3\frac{1}{2}$ in., or placing A at the mark for 5 in., in which case B would lie at the mark for 6 in. Supposing that we had not known the length of $\overline{AB}$, any one of the three examples just given would have told us that its length was 1 in., since 1 in. $= 2$ in. $- 1$ in. $= 3\frac{1}{2}$ in. $- 2\frac{1}{2}$ in. $= 6$ in. $- 5$ in. Because we were using an inch as our unit of measure, we would say that the measure of $\overline{AB}$ is 1.

Returning to the second case, we must ask what the measure of $\overline{AB}$ is in Fig. 4-6 (b), if its length is the same as in the other three figures, and how we can find our answer by relating to the real world. The answer is that we are using a "ruler" with a different unit of measure. If the length of $\overline{AB}$ were still 1 in., but we were using a foot-ruler, then wherever we put A, the point B would have a coordinate that was one-twelfth larger than the number under A; that is, the measure of $\overline{AB}$ would be $\frac{1}{12}$.

In other words, in the first three instances, we would say: "The length of $\overline{AB}$ is one inch"; in the other instance, we would say: "The length of $\overline{AB}$

is one-twelfth of a foot." It is well-known that one inch equals one-twelfth of a foot, and thus, either of the above statements would convey the same information.

If the length of $\overline{AB}$ were 1 in., but our ruler had as its unit of measure the centimeter, then the measure of $\overline{AB}$ would be approximately 2.54. Thus, while the length of a segment is fixed, the measure of the segment can vary and is seen to be completely dependent upon the unit of measure being used.

Now that we have discussed the matter in some detail, it is perhaps wise to clarify the difference between measure and measurement of some object. The **measure** of a set of points is simply a nonnegative number. The assignment of this number to a set depends upon the **unit of measure.** The unit of measure will be discussed at length in the subsequent paragraphs. The **measurement** of a set of points is a nonnegative number together with the name of the unit of measure.

EXAMPLES

(a) The length of $\overline{AM}$ is 13 in. "13" is the measure of $\overline{AM}$; the inch is the unit of measure; "13 inches" is the measurement of length of $\overline{AM}$.

(b) The length of $\overline{PQ}$ is 6.7 mm. "6.7" is the measure of $\overline{PQ}$; the millimeter is the unit of measure; "6.7 millimeters" is the measurement of length of $\overline{PQ}$.

How is the unit of measure determined in the abstract world of sets of points, and on the number line in particular? Many systems could be used; however, the one most widely accepted is to choose the distance on the number line between the origin and the point corresponding to the number "1" as the unit. The word "distance" is not quite correct here, since we are dealing with an abstract set of points. Yet, it has been chosen because it perhaps is the best one to convey the idea of unit of measure. The real heart of the idea is this: once the number "0" is assigned to some point of the line and the number "1" is assigned to a distinct second point of the line, then the assignment or correspondence of all other real numbers to all the other points of the line is completely determined. Then and only then can we know for certain the value of $b - a$, and consequently $m(\overline{AB})$.

One might ask how the decision is made as to which two points the assignment of 0 and 1 is given. The answer is that there is no rule or method. The assignment is completely arbitrary and can be made with any two points. These two points are given the name **unit pair.** Thus, when we speak of the measure of a segment in the abstract sense, we are always speaking relative to some unit pair. Furthermore, when we write $m(\overline{AB}) = m(\overline{CD})$ for two segments $\overline{AB}$ and $\overline{CD}$, it is always assumed that these measures have been derived relative to the same unit pair.

We should consider finding the measure of some segment $\overline{AB}$ when a unit pair, say M and N, have been chosen in such a way that $\overleftrightarrow{AB} \neq \overleftrightarrow{MN}$. Since we are thinking abstractly, we cannot really "move" $\overline{AB}$ so that it lies along the number line ruler determined by $\overline{MN}$. To handle this situation properly, it is necessary to introduce a new concept that is one of the most important in geometry. This is the idea of **congruent figures.** In an intuitive way, we can say that congruent figures have the same shape and the same size. Of course, this concept can be made much more precise and shall be so done when we turn our attention to triangles. The symbol for "is congruent to" is "$\cong$". Since all segments have the same shape, it is easy to tell when two segments are congruent. We define segment congruence as follows:

DEFINITION 4.4

$\overline{AB} \cong \overline{CD}$ if and only if $m(\overline{AB}) = m(\overline{CD})$.

Notice that the relation of congruence of segments is an equivalence relation on the set of all segments. This follows from the fact that congruence of segments is defined in terms of equality of real numbers, which is itself an equivalence relation.

We shall use the idea of congruence of segments to explain how we can "move" segments in our geometry. It is quite reasonable to think of the existence in space of countless numbers of segments that are congruent to some given segment, say $\overline{JK}$. Now, if we have any particular ray, there are, in fact, an infinite number of segments of that ray that are congruent to $\overline{JK}$. However, if we pick the endpoint of that ray, then there is only *one* segment congruent to $\overline{JK}$ on that ray that has as its endpoint the endpoint of the ray. This particular idea is very fundamental to the development of modern geometry; therefore, let us state the idea more formally as an assumption. (It is not a theorem that can be proven but is an idea regarding what we think we know about congruence of segments and the implications of this knowledge.)

POSTULATE 4.2

Given an arbitrary segment $\overline{JK}$, there exists a point X on some ray $\overrightarrow{PQ}$ such that $\overline{PX} \cong \overline{JK}$.

The word "arbitrary" in this postulate means, essentially, that we can have a segment of any length we might desire. Therefore, the postulate allows the possibility of "constructing" a segment of any size we wish any-where we wish. In the proofs of some theorems in later chapters and in the chapter on geometric constructions, this freedom is a most desirable asset to the development of the subject.

Let us see how we can apply this assumption to solve the problem at hand; namely, that of finding the measure of a segment $\overline{AB}$ when we are given a unit pair M and N such that $\overleftrightarrow{AB} \neq \overleftrightarrow{MN}$. If M and N are the unit pair, then $\overleftrightarrow{MN}$ is thought of as the "ruler" with which we find the measure of $\overline{AB}$. The choice of points M and N as the unit pair gives a coordinatization of $\overleftrightarrow{MN}$. Using the above assumption, we can either "move" the segment $\overline{AB}$ to the "ruler" $\overleftrightarrow{MN}$ or we can move the "ruler" $\overleftrightarrow{MN}$ to the segment $\overline{AB}$. Let us examine each case.

"Moving" $\overline{AB}$ to the ruler, we find a point X on ray $\overrightarrow{MN}$ such that $\overline{MX} = \overline{AB}$. It is an easy matter to find the measure of $\overline{MX}$. Referring to Fig. 4-7, we see $m(\overline{MX}) = x - m$, which is $x - 0 = x$, since we have chosen M to be the origin of the number line $\overleftrightarrow{MN}$. Since the assumption tells us $\overline{MX} \cong \overline{AB}$, we conclude by Definition 4.3 that $m(\overline{AB}) = m(\overline{MX})$ $= x$. Moving the "ruler" to $\overline{AB}$, we can find a point Y on ray $\overrightarrow{AB}$ such that $\overline{AY} \cong \overline{MN}$. Since M and N are a unit pair, then having $\overline{AY} \cong \overline{MN}$ essentially makes A and Y a unit pair and thus imposes upon $\overleftrightarrow{AB}$ the same coordinate system possessed by $\overleftrightarrow{MN}$. Thus, we are able to find the measure

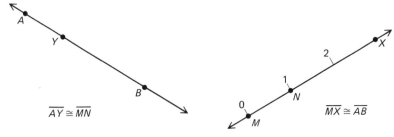

Figure 4-7

of $\overline{AB}$ relative to the given unit pair, M and N. It is important to note that making $\overline{AY} \cong \overline{MN}$ does not necessarily give A the coordinate zero. What it does do however, is give Y the coordinate $(a + 1)$, where a is the coordinate of A.

A noteworthy corollary to the preceding discussion is the fact that if a segment, $\overline{AY}$ in our example, is congruent to a segment $\overline{MN}$, where M and N are a unit pair, then A and Y may also serve as a unit pair in the sense that all measures found relative to A and Y will be the same as those found relative to the first pair, M and N.

Before passing to the measurement of angles, recall that when we defined the measure of $\overline{AB}$ as $m(\overline{AB}) = b - a$, we assumed that B was farther in the positive direction on the number line than A. In terms of the coordinates of A and B, we assumed $a < b$. In general, we cannot make this assumption, since we will not know the positions of A and B relative to the origin. Thus, the measure of $\overline{AB}$ must be redefined in order to take care of this situation. It is done in terms of absolute value which is defined below.

DEFINITION 4.5

$m(\overline{AB}) = |b - a|$.

DEFINITION 4.6

The **absolute value** of $b - a$, written as $|b - a|$, is defined as follows:

$|b - a| = b - a$ if and only if $b > a$

$|b - a| = - (b - a) = a - b$ if and only if $b < a$.

Because this definition helps to understand the concept of measure or length of a segment, many authors of geometry texts will use the absolute value symbol to denote the measure of a segment $\overline{AB}$ or the distance between A and B. Thus, you will see $|\overline{AB}|$ written in some texts. For us, this symbol is the same as $m(\overline{AB})$. This latter symbol will be used exclusively in this text.

EXERCISE SET 4.3

1. In the diagram below, assume that all named points are equally spaced apart. If Q and T are chosen as the unit pair such that the coordinate of Q is 0 and the coordinate of T is 1, what are the coordinates of A, M, N, V, P, R? What are the measures of $\overline{TA}, \overline{BR}, \overline{AV}, \overline{RQ}, \overline{QA}, \overline{RN}$?

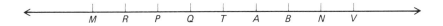

2. Repeat Exercise 1, where R and Q are chosen as the unit pair such that the coordinates of R and Q are 0 and 1, respectively.

3. Repeat Exercise 1, where A and P are chosen as the unit pair such that the coordinates of A and P are 0 and 1, respectively.

4. What can we say about the points A and B if $m(\overline{AB}) = 0$?

5. Distinguish between the length of a segment and the measure of a segment.

6. Suppose that A, B, and C are collinear and that the line on which they lie has a coordinate system so that the coordinates of A, B, and C are a, b, and c, respectively. Make up a definition for the betweenness of points in terms of the coordinates of those points.

7. Define the midpoint of a segment in terms of measure and betweenness.

8. Show how the textbook definition of absolute value dealing with number properties ($|x| = x$ if and only if $x \geq 0$; $|x| = -x$ if and only if $x < 0$) is simply a specific case of Definition 4.4. Relate the definition given in this exercise to the number line.

9. In the diagram below, all points are equally spaced. Evaluate the expressions on both sides of the blank and then decide whether the blank should be filled by $=$, $>$, $<$, or ?, where ? stands for indeterminate.

10. On the basis of the few examples in Exercise 9, try to make a generalization when the relation of equality will hold in such an example. Recall the definition of addition of cardinal numbers (or look it up in any text on the foundations of arithmetic) using the idea of cardinal number of a set. Is there any similarity between that definition and your generalization?

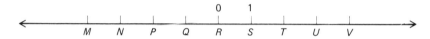

(a) $m(\overline{MP} \cup \overline{PQ})\_\_m(\overline{MP}) + m(\overline{PQ})$

(b) $m(\overline{NQ} \cup \overline{PT})\_\_m(\overline{NQ}) + m(\overline{PT})$

(c) $m(\overline{PR} \cup \overline{SV})\_\_m(\overline{PR}) + m(\overline{SV})$

(d) $m(\overline{NR} \cup \overline{PR})\_\_m(\overline{NR}) + m(\overline{PR})$

(e) $m(\overline{QS} \cup \overline{SU})\_\_m(\overline{QS}) + m(\overline{SU})$

(f) $m(\overline{MN} \cup \overline{NS})\_\_m(\overline{MN}) + m(\overline{NS})$

(g) $m(\overline{NR} \cup \overline{NS})\_\_m(\overline{NR}) + m(\overline{NS})$

(h) $m(\overline{QM} \cup \overline{QT})\_\_m(\overline{QM}) + m(\overline{QT})$

(i) $m(\overline{RP} \cup \overline{VQ})\_\_m(\overline{RP}) + m(\overline{VQ})$

(j) $m(\overline{PS} \cup \overline{MP})\_\_m(\overline{PS}) + m(\overline{MP})$

4.4 MEASURING AN ANGLE

The previous section has dealt with the measurement of segments. However, much of the development of the theory in that section is also relevant to the theory of measurement in general. To measure a segment, a real number is assigned to it on the basis of its size in comparison to the unit of measure. The unit of measure is also a segment, a geometric figure of the same type as that which was being measured. In the same way, when we seek the measure of an angle, we wish to assign to an angle some real number that will be called its measure; this number will be assigned on the basis of comparing a given angle to a unit of measure that will itself be an angle. Consequently, there are many similarities between measuring a segment and measuring an angle.

How does one measure an angle?   To relate to the measure of the sides of an angle would be fruitless—they are rays and, as such, extend infinitely far in one direction and therefore do not themselves possess a finite measure.   The same is true of the interior of an angle.   It is a set of points that is a subset of some plane—it, too, extends infinitely far.   The only way left to obtain the measure of an angle is to somehow measure the distance between the sides or to measure the amount of "opening" between the two sides.   This description is somewhat imprecise, and an attempt will be made to make the idea more clear.

Before doing so, we shall give a physical description.   Imagine two rays, $\overrightarrow{AB}$ and $\overrightarrow{AC}$ such that $\overrightarrow{AB} = \overrightarrow{AC}$.   (Recall two such rays were said to form a zero angle.)   Now, think of $\overrightarrow{AB}$ as remaining fixed and $\overrightarrow{AC}$ rotating so that $\overrightarrow{AB}$ and $\overrightarrow{AC}$ always remain in the same plane.   The two hands of a clock moving is a possible real analogue, although both hands of the clock move.   (This, incidentally, is a good physical model to use in an elementary classroom to get across the idea of an angle.)   At any rate, the amount of rotation of the one ray can be taken as a measure of the angle.   In some courses in mathematics, angles are thought of in this way—as a figure generated by a rotating side, $\overrightarrow{AC}$.   The point A is easily thought of as the center of a circle and C as a point moving along the circle.

It must be emphasized that, in this particular text, we do not consider movement and all points are considered fixed.   Nevertheless, thinking of a rotating ray does help to understand the idea of measurement of an angle.

Notice that congruence of angles has not yet been mentioned.   It is tempting to define congruence of angles in the same way we defined segments; that is, two angles are congruent if and only if they have the same measure. However, we have not yet shown how to measure angles, and as you will see in what follows, measurement of angles will be described in terms of congruent angles.   If this subject were being developed in a highly rigorous fashion, we would be in serious logical difficulty at this point, since we would be in danger of circularity in our definitions.   Even in our more informal development, we shall seek to avoid this circularity by assuming the idea of congruent angles to be sufficiently clear that we can take it as an undefined term.   We may describe two **congruent angles** by saying they have the same amount of opening between their sides or perhaps by saying that if it were possible to pick one of them up and place it directly on top of the other so that their vertices coincided, then so would their respective sides.

In segment measurement, there are several different units of measure that are in general usage.   Some of these are the inch, the foot, the yard, the meter, the centimeter; all are segments that have a different length.   With the angle, however, there is one generally accepted unit of measure—this is an angle that is described as having a size of one degree. The symbol for a degree is "°." In all ensuing discussion, we shall assume that any measures of angles have been derived relative to this standard unit of measure. Thus, if we write $m(\angle ABC) = 35$, we may rightfully infer the size of $\angle ABC$ is 35°.

What then is an angle of one degree? Consider a straight angle $\angle MNP$ with the vertex at N. Since the straight angle, $\angle MNP$, is the same set of points as a straight line, this set of points determines for us a half-plane. Imagine 179 rays, all having the same endpoint N and all lying in the same half-plane in such a manner that they form 180 small adjacent angles, each congruent to all the others and such that the interior of each small angle is disjoint from the interior of all the rest. All 180 of these angles have the same size and in a special way have partitioned the half-plane referred to into 180 congruent parts. Such an angle is an angle of one degree. The choice of 180 is due to ancient historical reasons having to do with the number of days in a year.

We are now ready to measure angles. We say:

1. The measure of a zero angle is 0;
2. The measure of a straight angle is 180;
3. All other angles have a measure m that is a real number with the property $0 < m < 180$.

Part (3) needs some amplification. This is done in the following postulate, which essentially states that the interior of any angle is a proper subset of some half-plane.

POSTULATE 4.3

Given any angle, $\angle ABC$, and some line, $\overleftrightarrow{XY}$, and a point P not on $\overleftrightarrow{XY}$, there is a point G on the P-side of $\overleftrightarrow{XY}$ such that $\angle GXY \cong \angle ABC$.

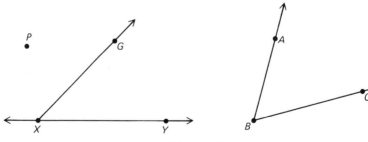

Figure 4-8

     Notice that this postulate is analogous to Postulate 4.2. This allows us to "construct" an angle of any size we desire anywhere we please by using any line as a starting place with any point on that line as the vertex.

     The half-plane described in Postulate 4.3 has a special property. If we consider the boundary line $\overleftrightarrow{XY}$ and some point M on $\overleftrightarrow{XY}$ that lies between X and Y, then the half-plane contains the totality of all rays that, together with ray $\overrightarrow{MX}$, form angles of all possible measures between 0 and 180. If this postulate is to help us in measuring angles, then it is necessary to impose a coordinate system on this half-plane in some way. Figure 4-9

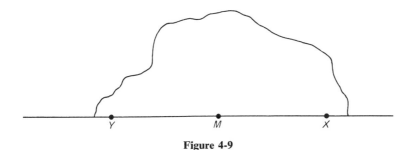

Figure 4-9

shows one way of doing this. Pictured is a continuously drawn curve with the property that any ray drawn with M as an endpoint will have an intersection of exactly one point with the curve. The ray drawn from M to this point together with $\overrightarrow{MX}$ forms an angle with measure less than 180. For each distinct point of the curve, we get a distinct angle. All such angles have $\overrightarrow{MX}$ as a side. The other sides of these angles are the totality of rays that fill up the half-plane. Since all the distinct angles have a different measure, we can assign a one-to-one correspondence between the points of this curved line and the real numbers between 0 and 180.

     This coordination of the plane is indeed special in that each point of the half-plane can be accounted for in some way. Each point lies on some ray that, together with $\overrightarrow{MX}$, forms an angle of unique measure. Thus, each point is assigned a unique real number between 0 and 180. However, the converse is not true. Each real number does not have a unique point of the half-plane assigned to it, but rather an infinity of points, these points all lying on the same ray.

     Figure 4-9 suggests a manner of constructing a measuring device for the measurement of angles. By placing $\overrightarrow{MX}$ so that it coincides with one of the sides of an angle and by making the vertices coincide, all one would have to do to get the measure of the angle is read the coordinate of the point on the curved line where the other side of the angle intersects the curved line. You are probably aware that such an instrument is in common usage. It is the familiar protractor, which is pictured in Fig. 4-10. The curved line of the protractor is a special curve, a semi-circle. Our discussion points out that such a special curve is not essential, but it is desirable in order to make reading the instrument easier.

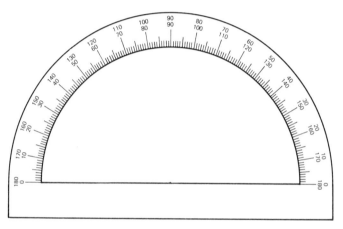

Figure 4-10

Since the measure of an angle is a positive real number, all statements regarding equality and inequality of the size of angles can be given meaning by relating to the measure of the angles. Certain angles are given special names because of their particular size.

DEFINITION 4.7

An angle is called a **right angle** if and only if its measure is 90.

DEFINITION 4.8

An angle is called an **acute angle** if and only if measure is less than 90 and more than 0.

DEFINITION 4.9

An angle is called an **obtuse angle** if and only if its measure is greater than 90 and less than 180.

DEFINITION 4.10

Two lines (or segments or rays thereof) are said to be **perpendicular** if and only if they intersect to form a right angle.

In all subsequent drawings where it is intended to convey the meaning that two lines are perpendicular, the symbol "$\neg$" will be used, as shown in Fig. 4-11. The symbol for perpendicular in text writing is "$\perp$." Thus, in Fig. 4-11, $\overline{AB} \perp \overline{CD}$.

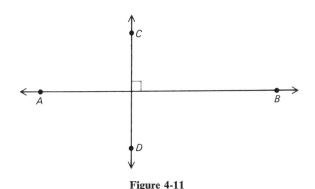

Figure 4-11

DEFINITION 4.11

Two angles are said to be **complementary** angles if and only if the sum of their measures is 90.

DEFINITION 4.12

Two angles are said to be **supplementary** angles if and only if the sum of their measures is 180.

Nothing in Definitions 4.11 or 4.12 implies that the two angles have to be adjacent. If two angles are complementary angles, each is said to be the **complement** of the other; if they are supplementary angles, each is said to be the **supplement** of the other.

EXERCISE SET 4.4

1. For each of the following angles, give the measure of its complement and its supplement:

   (a) $m(\angle ABC) = 30$ (b) $m(\angle MNP) = 47\frac{1}{2}$

(c) $m(\angle XYZ) = 34.9$ (d) $m(\angle GHJ) = (x - 120)$

(e) $m(\angle TVA) = (20\sqrt{2} - 5x)$

2. For parts (d) and (e) of Exercise 1, determine for what values of x the named angles will have a complement and a supplement.

3. Using a straightedge, draw angles with measures that you estimate to be 30, 45, and 60. Then use a protractor to see how close your estimates are.

4. Use a protractor to measure the angles in the pictured triangles. What is the sum of the measures of the three angles in each triangle?

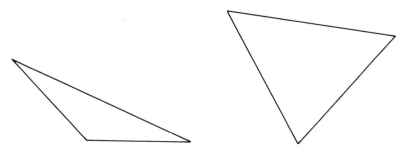

5. It is necessary to superimpose the straight line at the bottom of the protractor on one of the sides of an angle in order to find the measure of the angle? Explain your answer.

6. Given the configuration pictured below in which the number by each ray is the number associated with the angle formed by that ray and the ray $\overrightarrow{PQ}$. Find the measure of the named angles and tell whether each is acute, right, or obtuse.

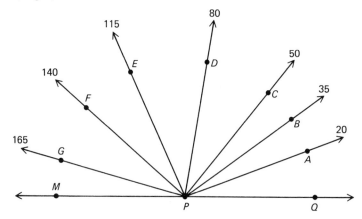

(a) $\angle APD$　　　　(d) $\angle GPD$　　　　(g) $\angle GPA$

(b) $\angle BPE$　　　　(e) $\angle FPC$　　　　(h) $\angle MPB$

(c) $\angle CPM$　　　　(f) $\angle DPF$

7. Is it necessary for the curved line of our angle measuring instrument to be drawn as a continuous curve? Explain your answer.

8. Is it true that a linear pair of angles are supplementary? Prove your answer.

9. What is true about two angles that are complementary to the same angle? Is it true if they are supplementary to the same angle? Can you prove it?

10. On the basis of the results of Exercises 8 and 9, prove that two vertical angles have the same measure.

11. Define the bisector of an angle.

12. Consult some well-known reference book on geometry or the history of mathematics to find the reason why a straight angle is subdivided into 180 congruent angles to get the unit of measure for an angle.

4.5　PRECISION AND ERROR IN MEASUREMENT

In the preceding two sections, we discussed the theory of measurement of geometric sets of points. We noted that there is no error in this type of measurement and that all measures are exact. In the measurement of real objects, however, the measure obtained is never exact. There is always some error of measurement involved. For any object, there is an exact measure belonging to it at any given moment, but we are unable to find it. It is true that the exact measure of an object may vary due to changes in temperature, pressure, or physical makeup, yet at a given moment the measure is only one number. Our inability to find this number exists for a variety of reasons; physical limitations of the measuring instrument, faulty use, or faulty reading of the instrument by the person employing it are perhaps the most common difficulties. Besides, the density property of real numbers tells us that between any two numbers, no matter how close, there are infinitely many other numbers; the exact measure could be any one of these. Common sense tells us that it would be downright impossible for the human eye or a machine, aided by the most powerful microscope imaginable, to distinguish such a number from the infinitely many others that are so close to it.

The science of measurement is not seriously impaired by this fact of inexact measurements. The solution is to get measurements that are as close to the exact measurement as the situation demands and as the precision of the measuring instruments, properly used, allows. The purpose of this section is to introduce some of the terminology used in the science of measurement. The examples used will relate to the measurement of segments, since it is assumed that this is the material with which you are most familiar. Some of the consequences of calculating with approximate data will be postponed until later chapters dealing with area and volume of geometric figures.

Suppose we wish to measure segment $\overline{MN}$, which is pictured in Fig. 4-12.

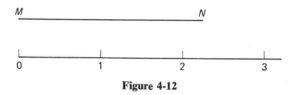

Figure 4-12

We have at our disposal a ruler that is calibrated in inches only. Placing the endpoint M over the point marked 0, we would read the measurement of $\overline{MN}$ as 2 in. Such a measurement implies that "2" is the measure of $\overline{MN}$ with reference to the inch as the unit of measure. This measurement also carries the implication that $\overline{MN}$ is 2 inches in length, correct to the nearest inch. "Correct to the nearest inch" means that the measurement of $\overline{MN}$ is closer to 2 in. than it is to 1 in. or to 3 in. Closer to 2 in. than to 1 in. means that $\overline{MN}$ is at least longer than $1\frac{1}{2}$ in.; closer to 2 in. than to 3 in. means that $\overline{MN}$ is less than $2\frac{1}{2}$ in. in length. Thus, if we did not have the benefit of Fig. 4-12 but were simply told that the measurement of $\overline{MN}$ was 2 in., correct to the nearest inch, we could correctly say $1\frac{1}{2} \leq m(\overline{MN}) < 2\frac{1}{2}$. Such a measurement is not too precise, and we might ask how much error is possible in terms of reporting 2 in. as the measurement. Since $|2 - 1\frac{1}{2}| = |2 - 2\frac{1}{2}| = \frac{1}{2}$, we say that the greatest possible error is $\frac{1}{2}$ in.

Let us repeat this procedure, using a differently calibrated ruler to measure $\overline{MN}$. This time we shall use one calibrated in half-inches, as pictured in Fig. 4-13. Again, placing M over the zero mark, we read the

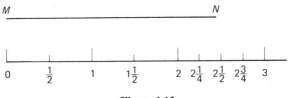

Figure 4-13

measurement of $\overline{MN}$ as $2\frac{1}{2}$ in., because the point N falls closer to the $2\frac{1}{2}$ in. mark than it does to the $2\frac{0}{2}$ in. mark or the $3\frac{0}{2}$ in. mark. Please note the usage of the "$\frac{0}{2}$" notation. If N had fallen closer to the 2 in. mark, we would have reported the measurement as $2\frac{0}{2}$ in. to denote that this measurement is correct to the nearest *half-inch*. If it had been reported as simply 2 in., then one would be unable to tell how closely the measurement was taken. In Fig. 4-13, the interval from $2\frac{1}{4}$ to $2\frac{3}{4}$ is marked off; if N had fallen anywhere in that interval, $m(\overline{MN})$ would have been reported as $2\frac{1}{2}$. In other words, reporting $m(\overline{MN})$ as $2\frac{1}{2}$ tells the reader that $2\frac{1}{4} \leq m(\overline{MN}) < 2\frac{3}{4}$. What is the greatest possible error in this case? Since $|2\frac{1}{2} - 2\frac{1}{4}| = |2\frac{1}{2} - 2\frac{3}{4}| = \frac{1}{4}$, the greatest possible error between reported measurement and actual measurement is $\frac{1}{4}$ in.

We shall very briefly repeat this measuring experiment, this time using a ruler marked off in quarter-inches. Refer to Fig. 4-14. Here N appears

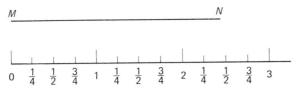

Figure 4-14

to be closer to the $2\frac{1}{2}$ mark than it is to the $2\frac{1}{4}$ mark. (The word "appears" is used deliberately to point out that the closer together the markings or calibrations of a ruler get, the more difficult it gets for the human eye to detect which mark the endpoint of a segment is closest to.) The measurement of $\overline{MN}$ would *not* be reported as $2\frac{1}{2}$ in. but rather as $2\frac{2}{4}$ in. to indicate that the measurement is correct to the nearest quarter-inch. Since $2\frac{2}{4}$ in. would be the reported measurement for all measurements between $2\frac{3}{8}$ in. and $2\frac{5}{8}$ in. if we were measuring correct to the nearest quarter-inch, the greatest possible error in this example would be $\frac{1}{8}$ in.

Certainly the last example gives the closest approximation to the actual measurement of $\overline{MN}$, whatever that measurement may be, since the last measurement had the smallest possible error, less than $\frac{1}{8}$ in. Of the three, it is said to be the most precise. The units—an inch, a half-inch, and a quarter-inch—that were used in the calibration of our different rulers are called **precision units**.

DEFINITION 4.13

Of two measurements of the same object, one is more **precise** than the second if it uses a smaller precision unit.

The **greatest possible error** (abbreviated g.p.e.) is easily computed to be one-half of the precision unit.

EXAMPLES

What is the precision unit and g.p.e. in the following reported measurements?

(a) $5\frac{7}{8}$ in.: The precision unit is $\frac{1}{8}$ in.

The g.p.e. is $\frac{1}{2}$ $(\frac{1}{8}) = \frac{1}{16}$ in.

(b) 14.063 ft.: The precision unit is 0.001 ft.

The g.p.e. is $\frac{1}{2}(0.001) = 0.0005$ ft.

One point should be noted. In the previous three cases related to Figs. 4-12, 4-13, 4-14, the discussion of the amount of error centered about the relative placement of the endpoint N of $\overline{MN}$ with respect to the markings of the measuring instrument. An assumption was made that point M was placed *exactly* above the zero point. Our discussion of measurement should tell us that this assumption cannot be made. Therefore, we should expect some difficulty in deciding just how accurate our measurements are.

Recall that we used the expression $2\frac{2}{4}$ rather than $2\frac{1}{2}$, because of the different meanings that each conveyed when taken in the context of measurement. Also recall that we referred to the measurement of $2\frac{0}{2}$ in. as having a meaning different from 2 in. These two instances are sufficient to give warning that ordinary rules of equivalence of fractions do not hold when dealing with measurement.

To make this clearer, let us consider the following example that frequently comes up in the classroom. Suppose a teacher asks the student to carry out the following subtraction:

$$2.7$$
$$-1.96$$
$$\overline{}$$

The typical advice is to add a zero following the digit 7 in the top row so as to make the number of digits in each number the same following the decimal point.   The justification is made that it makes no difference if terminal zeroes are added following a decimal point.   It is true that $\frac{7}{10} = \frac{7}{100}$ if we are dealing with exact numbers; however, if we are using numbers that have been arrived at from measurement, then 2.7 does not have the same meaning as 2.70, and one cannot necessarily be substituted for the other.   The numeral 2.7 carries the implication that the measurement is correct to the nearest tenth, which happens to be the seventh tenth, whereas 2.70 has the implication that the measurement is correct to the nearest hundredth, which happens to be the seventieth hundredth.   The terminal zero in 2.70 is said to be a significant digit; that is, the zero carries a meaning telling the reader the precision unit used was one one-hundredth of the unit of measure.

In measurement, a digit is significant if it gives some information about the precision unit used and also about the number of precision units contained in the measurement.   All nonzero digits are significant, as are zeroes occurring between two significant digits regardless of placement of the decimal point.   Thus, only initial and terminal zeroes are open to question. A terminal zero following a decimal point and other significant digits is significant; an initial zero following a decimal point but preceding a significant digit is not significant.   All terminal zeroes preceding a decimal point are not significant.

EXAMPLES

(a) 803 ft.
     All three digits are significant.

(b) 0.070 in.
     The first zero and the zero between the decimal point and the 7 are not significant; the 7 and the final zero are significant.

(c) 93,000,000 miles

     Only the 9 and the 3 are significant.

Example (c), which is the usual figure given for the distance from the earth to the sun, leads us to ask what precision unit is used in this measurement. Is it a million miles, ten thousand miles, a hundred miles, ten miles, or one mile? Unless further information is given in an example of this type, we must be careful not to assume the precision of measurement to be too great. Thus, we assume the largest possible, which is a million miles. The g.p.e. in this case is 500,000 miles—that is, we may be off in our measurement by a half-million miles! Is this a serious error? Is it as serious as being off a half-inch in a measurement of 3 in.?

The answer to the above question leads to the formalization of another term often used in measurement, that of relative error.

DEFINITION 4.14

The **relative error** of a measurement is the ratio of the g.p.e. of the measurement to the measurement itself.

The relative error is a simple number not connected to some unit of measure. Often, for ease of interpretation, it is expressed as a percent, in which case it is called the **percent of error**. To answer the questions at the end of the preceding paragraph, the relative error in measuring the distance to the sun is

$$\frac{500,000}{93,000,000} = 0.0054 = 0.54\%$$

Less than 1%. One-half inch in 3 inches is equal to $\frac{1}{2}/3 = \frac{1}{6} = 16.7\%$, which is rather sizeable.

There remains one last term to be examined—**accuracy**. In an intuitive way, we often think of it as being synonymous with "correctness." Mathematically, accuracy is directly related to relative error in that we say the measurement with the smallest relative error is the most accurate.

Generally this is true, but consider the following fictitious event. Suppose there is a segment whose *exact* length is 3.4007 in. Along come two students to measure this segment with their measuring instruments. The first, Able, has a rather crudely marked ruler that is correct only to the nearest quarter of an inch. Knowing the limitations of his ruler, Able carefuly places the ruler and reads the length as $3\frac{2}{4}$ in. With the above definition, it is easy to compute the relative error as $\frac{1}{8} \div 3\frac{2}{4}$, which equals $\frac{1}{28}$ or 3.6%. The second student, Baker, has a finely calibrated instrument that gives measurements correct to the nearest thousandth of an inch. Due to a scratch on the face of the ruler, he incorrectly reports the measurement as 13.401 in. rather than 3.401 in. Using the reported measurement, we can compute the relative error to be

$$(0.0005) \div 13.401, \quad \text{which equals} \quad \frac{1}{26802}, \quad \text{or} \quad 0.0036\%$$

If we went strictly by the rules, the second measurement would be far and away more accurate. Yet, armed with what we know about this example, such is not the case.

What is the lesson to be learned from the example? Basically, the rule regarding accuracy being related to relative error is true provided the measurement is a correct one. What is meant by "a correct one"? In a crude colloquialism, we could say that the reported measurement must "be in the same ballpark" with the exact measurement; that is, the two must be close together. Mathematically, it is this: the numerical or absolute value of the difference between the reported and exact measurements must be less than the g.p.e. Of course, one cannot guarantee that this is the case. All we can guarantee is that we try to learn how to properly use and manipulate an operative measuring instrument.

4.6 THE METRIC SYSTEM

As the earlier sections of this chapter noted, a person may choose any unit of measure for the basis of his measurements. However, for measurement to be useful in terms of communication to other persons, the standardized units of measure are to be preferred. The system of measurement used throughout most of the non-English speaking world is the metric system. In the United States, the metric system is legal and is used quite extensively in the scientific community. However, in the everyday usage we employ the English system with its inches, feet, yards, miles, pounds, ounces, pints, gallons, etc.

It is strongly recommended that the young schoolchild in the United States learn how to use the metric system. More and more Americans travel abroad each year to countries that use the metric system. The amount of foreign trade is increasing each year. Furthermore, there is the strong possibility that the United States will one day convert to the metric system in order to be more consistent with the rest of the world. If it is to carry out this change-over, a massive reeducation of the population must take place and the logical place for it to start is in the schoolrooms of the country.

The major advantage of the metric system, other than its almost universal usage, is the ease of conversion from one unit to another. The reason for this is that the metric system is basically a decimal system, and the conversion requires no more than the moving of a decimal point. The significant digits of a measure never change as they do in the English system. There is no need to remember all the different conversion factors that are necessary in the English system, as in converting from inches to yards, or pints to gallons, or feet to rods. Do you know, for example, how many feet make a rod or how many square feet make up an acre?

The purpose of this section is to introduce the terminology of the metric system and to serve as a reference for the prospective elementary teacher. All of the linear units in the English system are defined in terms of the basic linear unit of the metric system, the **meter.** For example, the meter is approximately equal to 39.37 in. But conversion from one system to the other is not recommended, as it can become a tedious problem in manipulation of large decimal numbers. Americans use the English system because it is familiar to them. A quart of liquid has meaning because they have seen and used a quart of milk or a quart of oil. A distance of five miles is meaningful because they have driven or walked such a distance many times. Such terms as liter, kilometer, kilogram have little meaning, simply because they have not used such terms. For example, if a person is 1.87 meters tall, is he short or tall? If we wish to fill up our gas tank at a filling station, will 20 liters be enough or will it take 30 liters or 40? If we still have 50 kilometers to go on a trip, is that far or "just around the corner," so to speak? Is a speed of 70 kilometers per hour excessive or is it "creeping" along?

The meter was originally defined as one ten-millionth of the distance from the equator to the north pole, measured along a meridian passing through Paris. It is officially defined as the length of a certain platinum-iridium bar that is kept at a certain temperature at the International Bureau of Weights and Measures at Sevres, France. Copies of this bar are kept in various centers around the world. Recently, the meter has been defined in terms of the wavelength of the orange-red light emitted by a lamp containing the chemical element, krypton 86. All of the other units of measure of length equal the meter times some integral power of ten. These are shown in Table 4-1.

TABLE 4-1

1 kilometer (km) $= 1000$ meters $= 1$ meter $\times 10^3$
1 hectometer (hm) $= 100$ meters $= 1$ meter $\times 10^2$
1 dekameter (dkm) $= 10$ meters $= 1$ meter $\times 10^1$
1 meter (m) $= 1$ meter $= 1$ meter $\times 10^0$
1 decimeter (dm) $= 0.1$ meter $= 1$ meter $\times 10^{-1}$
1 centimeter (cm) $= 0.01$ meter $= 1$ meter $\times 10^{-2}$
1 millimeter (mm) $= 0.001$ meter $= 1$ meter $\times 10^{-3}$

Each entry in the table is ten times as large as the entry immediately below it; for example, 1 cm = 10 mm. The prefixes are important, and the student should memorize them and their meanings, since these prefixes are also used in the measure of weight, where the basic unit is the **gram,** or in liquid measure, where the basic unit is the **liter.** For example, a kilogram is one thousand grams, a deciliter is one-tenth of a liter, a centigram is one one-hundredth of a gram, etc.

EXAMPLES

(a) Convert 3.47 m to cm.
As in the English system, going from a larger unit to a smaller, we must multiply by the conversion factor. Here it is $10^2 = 100$, so we multiply by 100, which is equivalent to moving the decimal two places to the right.
Ans.: 3.47 m = 347 cm.

(b) Convert 24.89 mm to m.
Going from smaller to larger, we divide by the conversion factor, which is $10^{+3} = 1000$. This division is equivalent to moving the decimal three places to the left.
Ans.: 24.89 mm = 0.02489 m.

(c) Convert 0.24 km to dm.
Larger to smaller, so multiply by a conversion factor of 10^1. This multiplication is the same as moving the decimal four places to the right.
Ans.: 0.24 km = 2400 dm.

As one learns the system, the conversion can be done quite mechanically by remembering that if you go up from one unit to the other (as they are listed in Table 4-1), then you move the decimal to the left the number of places that you have moved up; if you go down in the table, you move the decimal to the right the number of places that you move. It is very much the same as place value in a numeration system, and perhaps a horizontal array might make things easier to remember. The horizontal array is shown below.

10^3	10^2	10^1	10^0	10^{-1}	10^{-2}	10^{-3}
kilo-	hecto-	deka-	unit of measure	deci-	centi-	milli-

For purposes of reference, some of the more common conversion factors between the English and metric systems are given. The symbol "$\doteq$" means "approximately equal to."

$$1 \text{ inch} \doteq 2.54 \text{ cm} \qquad 1 \text{ liter} \doteq 1.06 \text{ qt}$$

$$1 \text{ meter} \doteq 39.37 \text{ in.} \qquad 1 \text{ kilogram} \doteq 2.2 \text{ lb}$$

$$1 \text{ mile} \doteq 1.61 \text{ km}$$

These are useful approximations that will give the reader some frame of reference. Figure 4-15 shows segment $\overline{AB}$, which is 10 cm in length.

A ── B

Figure 4-15

EXERCISE SET 4.5

1. Complete the following table:

Measurement	Precision unit	g.p.e.	Relative error
3⅝ in.			
3.060 ft			
2400 m			
240.0 m			
0.003 mm			

2. For the following measurements, give the range of measurements for which the given measurement would be the reported measurement:

(a) $7\frac{1}{2}$ in.   (e) 2.6 dm
(b) $2\frac{1}{16}$ yd   (f) 1.030 dm
(c) $4\frac{1}{4}$ ft   (g) 6 km
(d) 6.83 in.   (h) 2.43 cm

3. Are the limits of the ranges asked for in Exercise 2 exact measurements or are they approximate?

4. For each of the parts in Exercise 2, give the greatest possible error.

5. What is the percent of error for each of the parts of Exercise 2?

6. In the following measurements, tell which of the digits are significant:

(a) 560 miles   (e) 205,000 miles
(b) 0.003 cm   (f) 14.70 in.
(c) 94.07 mm   (g) 0.0403 mm
(d) 2,003 yd   (h) 26.8000 dm

7. Convert each of the metric measurements given below to the required measurement:

(a) 260.3 m to cm   (e) 26 km to cm
(b) 0.003 cm to dm   (f) 58.34 hm to dm
(c) 14.60 km to m   (g) 19.345 dkm to m
(d) 0.667 mm to dkm   (h) 0.14 m to mm

8. Suppose that you were called in as a consultant to help plan the re-education of a large number of people from the use of the English system to use of the metric system. What would you advise?

9. How would you go about educating young children about the metric system?

10. Suppose you were given a segment to measure and had a ruler that was calibrated in hundredths of an inch. If you used the mark at 2 in. as your zero point (that is, you placed that mark at one end of the segment), how much error would be introduced into your final measurement—that is, what is the greatest possible error?

11. Using the same measuring instrument as in Exercise 10, is there any other way of handling the instrument to reduce the size of the error?

SUGGESTED READINGS

[8] Ch. 8

[9] Ch. 5

[13] Ch. 1

[22] Ch. 4, 5, 6

[26] Ch. 3

[27] Ch. 5, 6, 7, 8

5

CONGRUENCE

Congruent figures have been described as those that have the same shape and size. The ideas regarding congruence of segments and angles will be extended to a larger class of geometric figures known as polygons. However, rather than use an informal, intuitive approach, the development will continue along the framework of a gradual shift in emphasis toward a more formal, precise approach. As noted in the introductory chapter, there are two sides of a mathematical development, the informal, intuitive approach and the formal, deductive one. They are not entirely mutually exclusive, yet the introduction to the formal side should be gradual. Thus, certain formal proofs will be introduced in the present chapter. Before we come to that portion, a section on a most important geometric figure, the simple closed curve will be given.

5.1 SIMPLE CLOSED CURVES

Intuitively, the idea of a curve is clear. To give it a precise, mathematical definition is not an easy task. Yet, its meaning should be understood, since, within the domain of the plane, it is perhaps the most basic geometric figure.

Although many persons feel that a curve must have "bends" or "wiggles" in it, the set of curves includes among its subsets, the sets of lines, segments, rays, and, as will be shown, polygons. One way to think of a curve is to imagine a picture of it being an object that could be drawn by a pencil moving along a piece of paper *without lifting the pencil from the page and without retracing any part*, except for a possible crossover point. This idea is not perfect, by any means, since we could not draw a picture of a line but, as best, only a half-line. (Even that might be a difficult task since a half-line extends infinitely far in one direction!)

If the basic unit of geometry—the point—were movable rather than an idealization in our minds, then its movement along some path would trace out a **curve**.[1] The fundamental mathematical concept here is that which may best be described as **continuity**. Continuity means that if a point were to "move" along a curve, there would always be a unique "next" point to "move" to until it came to an endpoint, if such an endpoint existed. The proper study of continuity is appropriate for a course in calculus.

Figure 5-1(a) shows some pictures of geometric figures that are curves, while Fig. 5-1(b) shows some that are not. Those in Fig. 5-1(b) could be classified as the union of curves however.

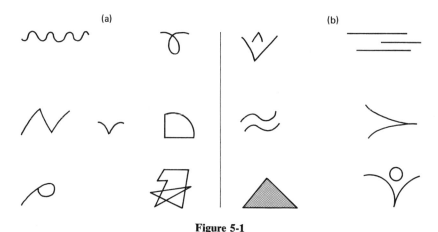

Figure 5-1

If a curve is such that it completely encloses a subset of the plane, then it is called a **closed curve**. This is saying that the beginning point of the path is the same as the endpoint, and, if a person were to look at a

---

[1] Some authors prefer to use the word "path" where the word "curve" is being used here, reserving the word "curve" for that subset of "path" consisting of nonstraight objects.

picture of a closed curve, he would be unable to tell where the beginning and end were. If the closed curve is such that it does not "cross over itself," then it is a **simple closed curve**. Figure 5-2(a) pictures some closed curves that are not simple, while Fig. 5-2(b) pictures only simple closed curves.

(a)   (b)

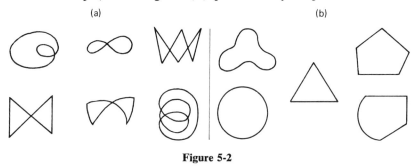

Figure 5-2

Just for fun, have you ever tried to draw the picture in Fig. 5-3 without lifting the pencil from the page? If it can be drawn, then what kind of a curve does it represent? If it cannot be drawn without lifting the pencil, then is it a picture of a curve at all?

Figure 5-3

Another interesting way to think of a simple closed curve is to think of the most "perfect" example of a simple closed curve, the circle. Suppose you took a piece of string attached in such a way that made it continuous and closed and laid it out on a table in a circular shape. If you then moved, or pushed, or nudged the string to transform its shape with the one stipulation that the string could not cross itself or even touch itself, then you would get representations of all kinds of simple closed curves. As elementary as this seems, these transformations of the circle, when viewed from a strictly mathematical point of view, have great theoretical interest for mathematicians. Such interest falls in the field of mathematics known as **topology**, which is sometimes called the "rubber sheet geometry." Drawing a circle on a rubber sheet and then pulling or stretching the rubber sheet without tearing it would give you many interesting variations of simple closed curves.

In some way, then, all simple closed curves have something in common and are related to the circle. One thing that they have in common is that they all completely enclose a subset of the plane in such a manner that one could travel from any point in that bounded region to any other point along a finite path of segments such that no segment of the path would intersect the curve. Thus, a simple closed curve partitions a plane into three mutually disjoint subsets, the curve itself, the interior of the curve, and the exterior of the curve.

DEFINITION 5.1

The union of a simple closed curve and its interior is called a **region.**

To get from the interior of a simple closed curve to its exterior, one has to cross the curve but once. However, there can be times when doing this is somewhat difficult—in fact, it may be hard to decide whether a given point is in the interior or the exterior, as Fig. 5-4 suggests. How would you decide whether P is in the exterior or interior?

Figure 5-4

5.2 POLYGONS

DEFINITION 5.2

A **polygon** is a simple closed curve consisting of the union of a finite number of segments.

It is worth emphasizing that a polygon consists only of those points that are the curve and not of those that we call its interior.

DEFINITION 5.3

The union of a polygon and its interior is called a **polygonal region.**

Polygons are usually classified according to the number of segments that make them up. The minimum number of **sides,** as the segments are called, is three. Table 5-1 gives some of the names of the more common polygons.

TABLE 5-1

No. of sides	Name
3	triangle
4	quadrilateral
5	pentagon
6	hexagon
7	heptagon
8	octagon
9	nonagon
10	decagon
n	n-gon

While a triangle is not a convex set (recall Definition 2.5), a triangular region is. However, when there are four or more sides to a polygon, the polygonal region associated with a polygon need not be convex. Figure 5-5

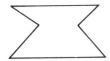 

Figure 5-5

illustrates several such cases, which are called **nonconvex** (or **concave**) polygons. Nonconvex polygons offer a certain amount of difficulty when considering the angles associated with these figures and, therefore, will receive limited treatment in this text. Unless otherwise stated, when the word "polygon" is used, it will mean a *convex* polygon. To say it another way, a convex polygon is one such that if any side is extended in both directions to produce a line, the rest of the polygon lies entirely in one half-plane determined by that line.

Much of the study of polygons can be reduced to a consideration of the triangle. Therefore, most of our attention will be directed to this familiar geometric figure.

DEFINITION 5.4

A **triangle** is a polygon with three sides.

It is important to note that the word "polygon" is necessary in the definition, since, without it, one could get figures such as those shown in Fig. 5-6. However, the word "angle" does not appear in the definition.

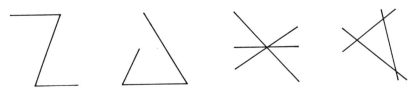

Figure 5-6

In fact, to be quite technical, a triangle does not have any angles, since the definition of angle requires the union of two rays. This deficiency is just a technicality. If the segments were extended in the necessary way, we would get the three angles associated with the triangle. For the sake of simplicity, these three angles are called the **angles of the triangle.** The vertices of these three angles are called the **vertices of the triangle,** and the triangle is named by means of these three points, with the symbolism $\triangle$. Thus, if the three vertices of a triangle were points A, B, and C, the triangle would be denoted by $\triangle ABC$. The order of listing the three vertices is unimportant.

Thus far, we have treated the interior of a polygon as an intuitively clear notion. Now, since we are armed with an understanding of the interior of an angle, it is a fairly easy matter to define the interior of a triangle.

DEFINITION 5.5

The **interior of a triangle** is the intersection of the interiors of the three angles of the triangle.

In one of the exercises you will be asked to define the interior of a polygon in terms of the interior of a triangle.

Triangles have special names that depend upon the relationships that exist between their sides and between their angles.

DEFINITION 5.6

A triangle is called **isosceles** if and only if at least two of its sides are congruent to each other.

DEFINITION 5.7

A triangle is called **equilateral** if and only if all three of its sides are congruent to each other.

DEFINITION 5.8

A triangle is called **scalene** if and only if no two of its sides are congruent to each other.

DEFINITION 5.9

A triangle is called **acute** if and only if all three of its angles are acute angles.

DEFINITION 5.10

A triangle is called a **right** triangle if and only if one of its angles is a right angle.

DEFINITION 5.11

A triangle is called **obtuse** if and only if one of its angles is an obtuse angle.

DEFINITION 5.12

A triangle is called **equiangular** if and only if all three of its angles are congruent to each other.

EXERCISE SET 5.1

1. Which of the figures pictured on page 92 are curves? Which are closed curves? Which are simple closed curves?

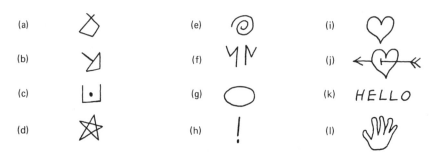

2. Suppose that it is possible to measure an "arc" of a curve, where an "**arc**" is defined to be a finite portion of the curve between two distinct points of the curve. Is it then possible to assign some finite real number as the measure of any curve? Of any closed curve? Of any simple closed curve?

3. An angle is an example of a curve. As we have seen, it is possible to describe the interior of an angle. Do all curves have an interior?

4. Are the following triples of numbers possible measures for the three sides of a triangle?

(a) 4, 7, 10   (e) 3, 4, 5
(b) 11.3, 5.9, 4.2   (f) 6.7, 6.7, 6.7
(c) 8, 8, 16   (g) 8, 15, 17
(d) 12, 5, 12   (h) 3.6, 7, 3.6

5. For those parts of Exercise 4 that you answered negatively, tell why you gave such an answer. For those that you answered affirmatively, further classify the triangles as to whether they are equilateral, isosceles, or scalene.

6. Draw at least five equilateral triangles. Measure the angles of the triangles. Is there any relationship evident?

7. Repeat the experiment of Exericise 6 with at least five isosceles triangles.

8. Do it again with at least five scalene triangles. (If you consider the size of the angles with respect to the length of the sides, there is a definite relationship here.)

9. May a triangle have two obtuse angles in it? Why? Try to answer this last part of the question without making use of the commonly known fact that the sum of the measures of the three angles of any triangle is 180.

10. It was earlier stated that we could get the three angles associated with

a triangle by extending the sides of the triangle in the "necessary" way. Describe in precise terms what is meant by the "necessary" way.

11. Define the interior of a polygon in terms of the interior of a triangle.

5.3 CONGRUENT TRIANGLES

We are now ready to consider the notion of congruent triangles. It is the intent to make the definition of congruent triangles mathematically precise and yet intuitively clear. Our intuitive notion is that if two congruent triangles were pictured on a piece of paper, we could cut out the resulting triangular regions and, by rotating and/or flipping them over, the two figures could be made to coincide exactly. They would be perfect matches of each other, somewhat akin to automobile components coming off an assembly line. All the parts of one triangle would match perfectly the corresponding parts of the other.

How can we make this idea precise? We shall begin with the mathematical idea that deals with this concept of matching, the notion of a one-to-one correspondence. Review Section 2.8 of the text at this point if the idea is not clear. It is necessary to set up a particular correspondence between two triangles that will guarantee that they have the same shape and size. Example (f) of Section 2.8 shows a one-to-one correspondence between two segments that are not the same size; thus, not any correspondence will do. Moreover, we need one that will also match up the angles of the two triangles correctly. As a matter of fact, the correspondence between the two triangles is described by means of the vertices of the triangles, since using vertices is the way that a triangle is named.

Perhaps it might be wise at this point to use a specific example to clarify this concept. Figure 5-7 shows two triangles, $\triangle ABC$ and $\triangle DEF$. If we were to consider only the sets $\{A, B, C\}$ and $\{D, E, F\}$ and the possible one-to-one correspondences that could exist between these two finite sets, we would conclude that there are six possible one-to-one correspondences. Three of these are shown. You should convince yourself that there are three more by listing them as well.

$$
\begin{array}{ccc}
A \leftrightarrow D & A \leftrightarrow E & A \leftrightarrow E \\
B \leftrightarrow E & B \leftrightarrow F & B \leftrightarrow D \\
C \leftrightarrow F & C \leftrightarrow D & C \leftrightarrow F
\end{array}
$$

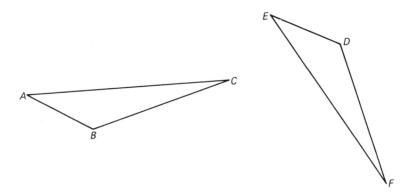

Figure 5-7

But, we do not consider only the matchings of vertices of two triangles; we must match up the sides as well. Thus, for two triangles that are to be matched up, there are six corresponding parts—three angles and three sides.

A convenient shorthand notation has been devised that indicates all six pairings. It is $ABC \leftrightarrow DEF$, which gives the pairings between the vertices

$A \leftrightarrow D$, $B \leftrightarrow E$, $C \leftrightarrow F$, and the pairings between the sides $\overline{AB} \leftrightarrow \overline{DE}$,

$\overline{BC} \leftrightarrow \overline{EF}$, $\overline{AC} \leftrightarrow \overline{DF}$. The pairings (or matchings) indicated are determined by the order in which the letters are listed. For example, $ABC \leftrightarrow DEF$ indicates a different correspondence than $ABC \leftrightarrow EFD$, since the latter indicates the six matchings of the corresponding parts of the triangles

as follows: $A \leftrightarrow E$, $B \leftrightarrow F$, $C \leftrightarrow D$ and $\overline{AB} \leftrightarrow \overline{EF}$, $\overline{BC} \leftrightarrow \overline{FD}$, $\overline{AC} \leftrightarrow \overline{ED}$. We are just about ready to give the definition to two congruent triangles. At the risk of being repetitive, we must emphasize that the matching up has to ensure that the triangles will have the *same size and shape*.

DEFINITION 5.13

Two triangles, $\triangle ABC$ and $\triangle DEF$, are **congruent** (written $\triangle ABC \cong \triangle DEF$) if and only if there exists a one-to-one correspondence between the vertices of the two triangles such that the six corresponding parts of the triangles are congruent.

We see that, after this long introduction, this idea is basically a simple one and depends upon a knowledge of what is meant by congruent segments and congruent angles.

Returning to Fig. 5-7, $\triangle ABC$ is congruent to $\triangle DEF$, since there is a correspondence that does give congruent corresponding parts; it is the correspondence $ABC \leftrightarrow EDF$. In order to achieve consistency of notation, the letters of the vertices used to name two triangles that are congruent will be listed in the order that indicates the correspondence that ensures the congruence in the first place. For example, if "$\triangle ABC \cong \triangle EDF$" is written, then this symbolism implies that $ABC \leftrightarrow EDF$ is a correspondence that gave the congruence; thus, a reader will immediately be able to tell which corresponding parts are congruent to each other. Using this same example, if we write $\triangle ABC \cong \triangle EDF$, then we know immediately as a consequence of this fact, and from the definition of congruent triangles, that

$$\angle A \cong \angle E, \quad \angle B \cong \angle D, \quad \angle C \cong \angle F, \quad \overline{AB} \cong \overline{ED}, \quad \overline{BC} \cong \overline{DF},$$

$$\overline{AC} \cong \overline{EF}$$

EXERCISE SET 5.2

1. In each of the parts of this exercise, a figure is given and a correspondence between two triangles is indicated. Fill in the blanks below each part.

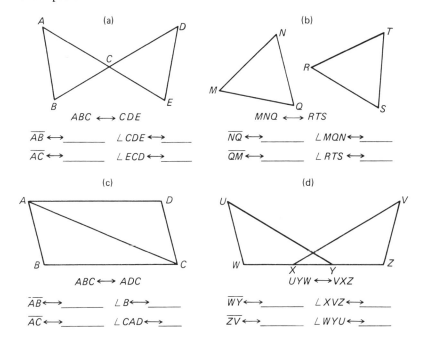

(a)

$ABC \longleftrightarrow CDE$

$\overline{AB} \longleftrightarrow$ _____ $\angle CDE \longleftrightarrow$ _____

$\overline{AC} \longleftrightarrow$ _____ $\angle ECD \longleftrightarrow$ _____

(b)

$MNQ \longleftrightarrow RTS$

$\overline{NQ} \longleftrightarrow$ _____ $\angle MQN \longleftrightarrow$ _____

$\overline{QM} \longleftrightarrow$ _____ $\angle RTS \longleftrightarrow$ _____

(c)

$ABC \longleftrightarrow ADC$

$\overline{AB} \longleftrightarrow$ _____ $\angle B \longleftrightarrow$ _____

$\overline{AC} \longleftrightarrow$ _____ $\angle CAD \longleftrightarrow$ _____

(d)

$UYW \longleftrightarrow VXZ$

$\overline{WY} \longleftrightarrow$ _____ $\angle XVZ \longleftrightarrow$ _____

$\overline{ZV} \longleftrightarrow$ _____ $\angle WYU \longleftrightarrow$ _____

2. Is it possible that more than one of the six possible one-to-one correspondences between the vertices of two triangles could ensure that two
   triangles are congruent?   Explain your answer.

3. Aside from the trivial one-to-one correspondence $ABC \leftrightarrow ABC$, is it
   possible that a triangle could be congruent to itself in another way?

4. Is the relation of triangle congruence an equivalence relation on the set
   of all triangles?   If not, tell which of the three properties it fails to
   satisfy.

5. Be prepared to discuss the following definition of congruent triangles.
   Two triangles are said to be congruent to each other if there exists a
   one-to-one correspondence between the points of these triangles, so that
   for any two pairs of corresponding points X and X' and Y and Y',

   $\overline{XY} \cong \overline{X'Y'}$.   This is pictured below.

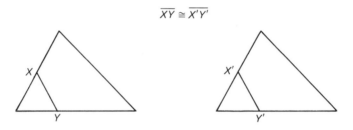

5.4   CONGRUENCE POSTULATES FOR TRIANGLES

Congruent triangles are of great use in the development of a study of geometry.   There will be many times when we shall want to know if two
triangles are congruent to each other.   The definition of congruence between
two triangles states under what conditions we may be sure that a congruence
exists between two triangles.   However, it is of interest, both from a theoretical point of view as well as a practical one, to know whether we can establish
a congruence, given less than the six corresponding congruent parts.   Putting
this another way, we can ask what minimum number of congruent corresponding parts is needed to ensure the existence of a congruence between
two triangles.   Usually this question can be answered by the number "3,"
although this answer must be qualified by saying that not just any three
parts will do—the three parts must be in certain positions relative to
each other.

      First, some terminology must be introduced.   If we look at $\triangle ABC$ in

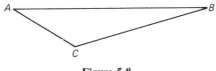

Figure 5-8

Fig. 5-8, side $\overline{BC}$ is said to be included between $\angle B$ and $\angle C$. Also, $\angle A$ is said to be included between sides $\overline{AB}$ and $\overline{AC}$.

At this point, you are advised to go back to Section 1.2 of the first chapter and reread the first two paragraphs of that section. The reason for this request is that several postulates about congruent triangles shall be introduced, and it is felt to be important that you appreciate the role of a postulate in the development of a mathematical system.

Our study of geometry is concerned with *relationships* that exist between geometric figures. Definitions and undefined terms get us started and tell what the figures are and what the relationships consist of. It then usually remains to be told by the theorems under what conditions these relationships hold. The proofs of the theorems rest upon a logical development that uses the definitions and previous theorems as its basis. However, to make progress with the theorems, it is necessary to state a few "theorems" that require no proof. This is completely analogous to having to use some undefined terms. We can keep going back to "primary" sources, but there comes a time when we must stop and state: "This is the beginning." As it is true with the undefined terms in the realm of definitions, so it is also true with our unproven theorems in the realm of theorems. These accepted-as-true statements are called **postulates.**

Often, the postulates deal with some very fundamental relationships that are obvious, but this is, by no means, always the situation. You must realize that these postulates, obvious or not, *have* to be made. As stated in Chapter 1, many mathematicians choose to try to limit these postulates to the very minimum. However, some choose to be a bit more practical and state some postulates that could be proven on the basis of given definitions and other axioms, and that, therefore, qualify as bona fide theorems. One reason for this choice is that the proofs tend to be very tedious and somewhat "fuzzy" to some students. As a result, the proofs could be confusing rather than enlightening. So it is with the postulates on triangle congruence that will be given in the following paragraphs. Euclid tried to prove them as theorems, but his classical proofs were very laborious and, in fact, subject to some question by modern students of geometry as being invalid. At any rate, it is hoped that the statements will be obvious; they would be if you were asked previously to perform many experiments involving measuring of sides and angles of pairs of triangles. Actually, in some of the exercises that follow, you will be asked to try several measuring experiments that should convince you that these postulates are reasonable and believable.

POSTULATE 5.1 (Side-Angle-Side)

Given a one-to-one correspondence between the vertices of two triangles, if two sides and the included angle of one triangle are congruent to the corresponding parts of the other, then the triangles are congruent to each other.

POSTULATE 5.2 (Angle-Side-Angle)

Given a one-to-one correspondence between the vertices of two triangles, if two angles and the included side of one triangle are congruent to the corresponding parts of the other, then the triangles are congruent to each other.

POSTULATE 5.3 (Side-Side-Side)

Given a one-to-one correspondence between the vertices of two triangles, if three sides of one triangle are congruent to the corresponding parts of the other, then the triangles are congruent to each other.

It is the SSS postulate that is most often proven as a theorem in other textbooks. Most high school geometry texts prove it as a theorem, and therefore the chances are good that you have seen a proof of this theorem at some time or other. What is of major concern is that you realize, whether these statements be called theorems or postulates, the importance of these statements and what they imply. The congruence of triangles may be inferred whenever the conditions of these statements have been met. Before we turn to some exercises, it should be noted in passing that nowhere in the previous discussion has the word "distinct" been used when referring to the two triangles.

EXERCISE SET 5.3

1. (This exercise is designed to show that the SAS postulate is reasonable.) Draw three triangles, $\triangle ABC, \triangle MNP,$ and $\triangle XYZ.$ Let them be of

any size or shape that you wish. Using a protractor, measure the angles
$\angle B$, $\angle N$, and $\angle Y$. Below your original triangles construct with the
protractor new angles, $\angle B'$, $\angle N'$, and $\angle Y'$, so that they are congruent
to the original angles. Then, on the rays that are the sides of these
new angles, mark off segments congruent respectively to the including
sides of the original angles, $\angle B$, $\angle N$, and $\angle Y$. Now draw segments
$\overline{A'C'}$, $\overline{M'N'}$, and $\overline{X'Y'}$, so that the triangles $\triangle A'B'C'$, $\triangle M'N'P'$, and
$\triangle X'Y'Z'$ are formed. Measure the three remaining pairs of correspond-
ing parts of the three paired sets of triangles to convince yourself that
they are congruent parts.

2. Repeat Exercise 1 for the ASA postulate by drawing three new triangles
   and choosing one side of each and its including angles to construct the
   new triangles upon which the comparison is to be made.

3. We wish to check on the SSS postulate as well. If you are given a
   randomly chosen $\triangle ABC$, how can you construct a copy of it, so that
   you can check on the congruence of the three pairs of corresponding
   angles? Once you get the answer, then repeat Exercise 1 for the SSS
   postulate.

4. What would be the SSA postulate? Why was it not stated?

5. What would be the AAA postulate? Why was it not stated?

   In the following exercises the parts of the triangles that are marked
alike are congruent.

6. All parts of this question refer to the following figures.

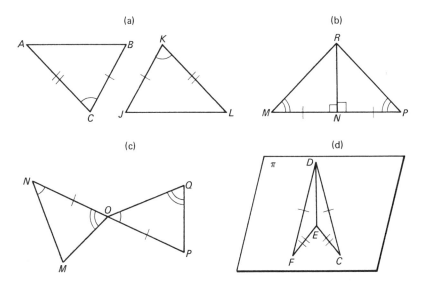

(a)   (b)

(c)   (d)

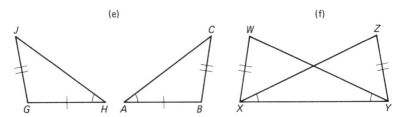

(a) In figure (a), what are the sides that include ∠K?

(b) In figure (a), what is the angle included by $\overline{JK}$ and $\overline{LJ}$?

(c) In figure (c), $\overline{QO}$ and $\overline{PO}$ include which angle?

(d) In figure (d), what are the including sides of ∠DEF?

(e) In figure (d), what are the sides that include ∠C?

(f) In figure (f), what are the including angles of $\overline{WY}$?

(g) In figure (f), $\overline{XY}$ is included by what two angles?

7. In which of the pictures shown in Exercise 6 can you prove by means of one of the congruence postulates that the pairs of triangles are congruent? Tell which postulate is used (SAS, ASA, or SSS) and indicate the correspondence that ensures the congruence.

5.5   SOME EASILY PROVEN THEOREMS

In this section, we shall set forth some theorems whose proofs follow rather easily from the congruence postulates and earlier definitions and from results of exercises. The first is a well-known theorem regarding the angles of an isosceles triangle. Recall that an isosceles triangle is one with two sides congruent to each other. The angle included by these two sides is called the **vertex angle** and the other two angles are called the **base angles** of the isosceles triangle. The theorem is often called the **Isosceles Triangle Theorem.**

THEOREM 5.4

The base angles of an isosceles triangle are congruent to each other.

Before showing the proof, we shall state it in another way, the "if-then" form: If two sides of a triangle are congruent to each other, then the angles opposite those sides are congruent to each other. The proof will refer to

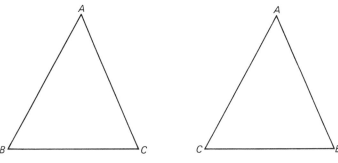

Figure 5-9

Fig. 5-9. This proof is an interesting one, since it uses the somewhat novel technique of proving a triangle congruent to itself. Recall Exercise 4 of Exercise Set 5.2 in which it was found that triangle congruence is an equivalence relation.

Proof. Consider the one-to-one correspondence $ABC \leftrightarrow ACB$, remembering that $\overline{AB} \cong \overline{AC}$.

Statement	Reason
1. $\overline{AB} \cong \overline{AC}$	1. Hypothesized as true
2. $\angle A \cong \angle A$	2. Reflexive property of angle congruence
3. $\overline{AC} \cong \overline{AB}$	3. Symmetric property of congruence of segments, using the congruence given in Statement 1
4. $\triangle ABC \cong \triangle ACB$	4. SAS Postulate, using the correspondence $ABC \leftrightarrow ACB$
5. $\angle B$ $\angle C$	5. Definition of triangle congruence

The next theorem is the converse of the Isosceles Triangle Theorem.

THEOREM 5.5

If two angles of a triangle are congruent, then the two sides opposite those angles are congruent to each other; that is, the triangle is isosceles.

Proof. The technique of this proof is the same as the one above; this time we shall again consider the correspondence $ABC \leftrightarrow ACB$, noting that the conditions of the statement of this theorem tell us to assume the truth of the statement $\angle B \cong \angle C$ (using Fig. 5-9 again).

Statement	Reason
1. $\angle B \cong \angle C$	1. Hypothesized as true
2. $\overline{BC} \cong \overline{CB}$	2. Reflexive property of congruence of segments
3. $\angle C \cong \angle B$	3. Symmetric property of angle congruence, using the congruence given in Statement 1
4. $\triangle ABC \cong \triangle ACB$	4. ASA Postulate ($ABC \leftrightarrow ACB$)
5. $AB \cong AC$	5. Definition of triangle congruence

Another familiar theorem is the one that follows. Its proof follows immediately and quite simply from the Isosceles Triangle Theorem. Theorems that follow directly and simply from another theorem are often called **corollaries** to the theorem.

COROLLARY 5.6

An equilateral triangle is also equiangular.

Proof. Refer to Fig. 5-10.

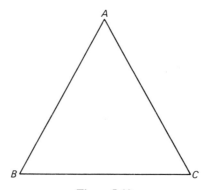

Figure 5-10

Statement	Reason
1. $\triangle ABC$ is equilateral	1. Hypothesized as true
2. $\overline{AB} \cong \overline{AC}$	2. Definition of equilateral triangle
3. $\angle C \cong \angle B$	3. Isosceles Triangle Theorem
4. $\overline{AB} \cong \overline{BC}$	4. Definition of equilateral triangle
$\angle C \cong \angle A$	5. Isosceles Triangle Theorem
6. $\angle A \cong \angle B \cong \angle C$	6. Transitivity property of angle congruence
7. $\triangle ABC$ is equiangular	7. Definition of equiangular triangle

The converse of this theorem could also be proven, using the converse of the Isosceles Triangle Theorem. However, its proof is almost obvious and will be omitted here.

One other theorem will be given. It contains several words that need definition: median, altitude. These words pertain to certain segments associated with triangles.

DEFINITION 5.14

A **median** of a triangle is a line segment, one of whose endpoints is the vertex of an angle of the triangle, while the other endpoint is the midpoint of the segment that is the opposite side of the triangle.

There are, of course, three medians in a triangle.

DEFINITION 5.15

The **altitude** of a triangle is a line segment, one of whose endpoints is a vertex of an angle of the triangle and whose other endpoint is on the line that contains the opposite side of the triangle with the property that it is perpendicular to that line.

This definition sounds a bit complicated. Let us illustrate it in Fig. 5-11, in which one of the altitudes of each triangle is shown by the dotted lines. There are three altitudes for every triangle.

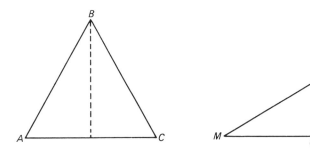

Figure 5-11

THEOREM 5.7

The median from the vertex angle of an isosceles triangle is also an altitude of that triangle.

The situation is pictured in Fig. 5-12. We are given that point D is the midpoint of $\overline{BC}$. We are required to prove that $\overline{AD} \perp \overline{BC}$. Thus far, our only method of showing two segment (or lines) to be perpendicular is

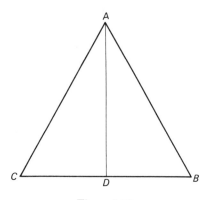

Figure 5-12

by the definition given in Chapter 4; namely, to show that $\angle ADC$ and $\angle ADB$ are right angles. How shall we show that these two angles are right angles? Certainly not by saying $\overline{AD} \perp \overline{BC}$! We would be guilty of circular reasoning if we did that. We *can* do it by showing that each of them has a measure of 90. We can see that they form a linear pair, and Exercise 8 of Exercise Set 4.4 assures us that they are also supplementary. Since we know that supplementary angles are two angles the sum of whose measures is 180, all we need do is to show that $\angle ADC$ and $\angle ADB$ have equal measures. How do we do that? That is the heart of the proof!

Proof. Consider Fig. 5-12 and the correspondence $ADC \leftrightarrow ADB$.

Statement	Reason
1. $\overline{AD}$ is a median of $\triangle ABC$.	1. Hypothesized as true
2. D is the midpoint of $\overline{BC}$.	2. Definition of median
3. $\overline{DC} \cong \overline{DB}$	3. Definition of midpoint of a segment
4. $\overline{AD} \cong \overline{AD}$	4. Reflexive property of congruence of segments
5. $\overline{AC} \cong \overline{AB}$	5. Hypothesized as true (the triangle is isosceles)
6. $\triangle ADC \cong \triangle ABC$	6. SSS Postulate
7. $\angle ADC \cong \angle ADB$	7. Definition of triangle congruence
8. $m(\angle ADC) = m(\angle ADB)$	8. $\angle ADC \cong \angle ADB$

The rest of the proof is omitted, since it has been outlined in the previous paragraph.

In some of the proofs in the exercises that follow, you will be dealing with the measure of segments and angles. Remember that these measures are real numbers and, as such, can be treated by the addition, multiplication, and cancellation properties for equalities of real numbers.

EXERCISE SET 5.4

1. Given the figure pictured below. If $\overline{AB} \cong \overline{CD}$ and $\triangle EBC$ is isosceles, then prove $\triangle EAD$ is also isosceles. Supply the missing reasons.

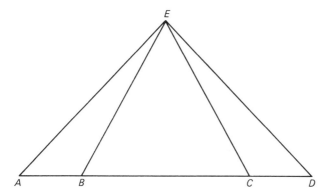

Proof: Consider the correspondence $EAB \leftrightarrow EDC$.

Statement	Reason
1. $\triangle EBC$ is isosceles, with $\overline{EB} \cong \overline{EC}$.	1. ?
2. $\angle EBC \cong \angle ECB$	2. ?
3. $\angle EBA \cong \angle ECD$	3. Supplements of congruent angles are congruent (Exercise 9 of Exercise Set 4.4).
4. $\overline{AB} \cong \overline{CD}$	4. Hypothesized as true
5. $\overline{EB} \cong \overline{EC}$	5. ?
6. $\triangle EAB \cong \triangle EDC$	6. ?
7. $\overline{EA} \cong \overline{ED}$	7. ?
8. Therefore, $\triangle EAD$ is isosceles	8. ?

2. Prove: The median from the vertex of an isosceles triangle also bisects

the vertex angle. (*Hint*: The first seven steps are the same as the last proof of the previous section. What is the definition of an angle bisector?)

3. Prove: The medians drawn from the base angles of an isosceles triangle are congruent.

   Proof: The hypothesized situation is pictured below. $\triangle ABC$ is isosceles, with $\overline{AB} \cong \overline{AC}$, and $\overline{BN}$ and $\overline{CM}$ medians. Consider the correspondence $BMC \leftrightarrow CNB$.

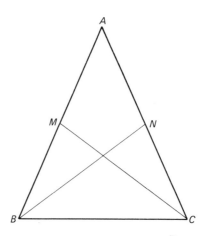

Statement	Reason
1. $\overline{AB} \cong \overline{AC}$	1. ?
2. $m(\overline{AB}) = m(\overline{AC})$	2. Definition of segment congruence
3. $\overline{BN}$ and $\overline{CM}$ are medians.	3. ?
4. $m(\overline{BM}) = \frac{1}{2}m(\overline{AB})$; $m(\overline{CN}) = \frac{1}{2}m(\overline{AC})$	4. ?
5. $m(\overline{AB}) = \frac{1}{2}m(\overline{AC})$	5. Multiplication property of equality; each side of the equation in Statement 2 was multiplied by $\frac{1}{2}$.
6. $m(\overline{BM}) = m(\overline{CN})$	6. Transitive property of equality of real numbers
7. $\overline{BM} \cong \overline{CN}$	7. ?
8. $\angle MBC \cong \angle NCB$	8. ?
9. $\overline{BC} \cong \overline{CB}$	9. ?
10. $\triangle BMC \cong \triangle CNB$	10. ?
11. $\overline{BN} \cong \overline{CM}$	11. ?

5.6 CONGRUENT POLYGONS

Now that we have ways of determining if triangles are congruent, it becomes an easy matter to decide if polygons of any number of sides are congruent to each other. Of course, it is necessary that the two polygons under consideration have the same number of sides. You may have already guessed that the method is to partition the polygon into triangles by drawing diagonals from corresponding vertices and then see if the resulting triangles are congruent in pairs.

Let us illustrate this method by means of an example. In Fig. 5-13 two pentagons (five-sided polygons) are shown with certain diagonals drawn.

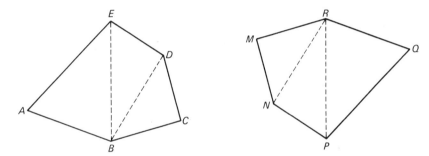

Figure 5-13

There are many (how many?) one-to-one correspondences that could be established between the two sets of five vertices. Rather than belabor the point, we state directly that the correspondence that works is $ABCDE \leftrightarrow QRMNP$, where the order of writing indicates the five pairs of corresponding vertices and five pairs of corresponding sides in a manner similar to what was done with triangles. The manner of determining whether or not a correspondence gives a congruence is to draw all possible diagonals from two corresponding vertices and then, if the resulting corresponding triangles are congruent in pairs, the polygons are congruent. In the example given, $\triangle ABE \cong \triangle QRP, \triangle EBD \cong \triangle PRN, \triangle DBC \cong \triangle NRM$. One would check for the triangle congruences by the methods suggested in this chapter. No formal postulates or theorems for congruent polygons will be stated here.

One more definition is in order before leaving this section on polygons.

DEFINITION 5.16

A **regular polygon** is one all of whose sides are congruent to each other and all of whose angles are congruent to each other.

In other words, a regular polygon is one that is both *equilateral* and *equiangular*. It must, of necessity, be a convex polygon—that is, its interior is a convex set. Some of the more familiar regular polygons are pictured in Fig. 5-14.

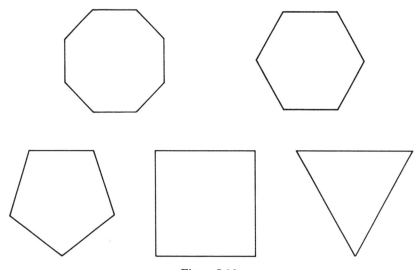

Figure 5-14

A word of caution! Except for the triangle, congruent sides of a polygon do not imply congruent angles, as several of the polygons illustrated in Fig. 5-15 on page 109 demonstrate. Each of the polygons illustrated is equilateral but not equiangular.

5.7   SYMMETRY

The preliminary intuitive discussions of congruence treated congruent figures as being those that could be placed one upon another in such a way that they fit exactly. These discussions centered on segments, angles, triangles, and polygons, all of which are sets of points that are subsets of a straight line or a union of such subsets. There is another property of a single set

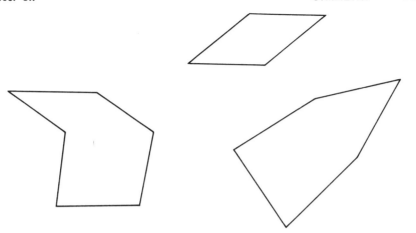

Figure 5-15

of points that is somewhat related to the intuitive notion of congruence. This is the idea of symmetry.

Most persons have an informal, yet perhaps strong, feeling for symmetry. We say that an object is symmetrical if it has the property that, in some way, it could be broken into two parts and these parts folded or rotated in some way so that they would fit exactly one upon the other. Figure 5-16 shows objects that possess this property of symmetry.

Figure 5-16

Since symmetry seems to be so closely related to congruence, can it be described or explained in terms of congruence? The answer is in the affirmative. The two matching parts of a symmetric figure can be thought of as sets of points in their own right, and, as such, can be considered to be congruent to each other. Congruence was taken as an intuitive idea in the first place—in fact, congruence for angles was an undefined term—and there is nothing to prevent the extension of this notion to all sorts of geometric figures.

It is possible to describe in a precise manner how the matching up of the two parts of a symmetric figure takes place. This description is made in two ways—symmetry about a line and symmetry about a point. Let us take each in turn.

Symmetry about a line deals with figures that have the property of appearing to be the result of someone's drawing a figure on a piece of paper with wet ink and then folding that paper down a straight line to get a "mirror image" or "reflection" on the other side of the fold. This straight line, which is the fold, is called an **axis of symmetry**.

DEFINITION 5.17

A figure is **symmetric about a line** l, called its axis of symmetry, if for all points A of the figure there exists a symmetric partner A' for A (not necessarily distinct), so that $\overline{AA'}$ is perpendicular to l at some point M and such that $\overline{AM} \cong \overline{A'M}$.

This definition is exemplified in the parts of Fig. 5-17.

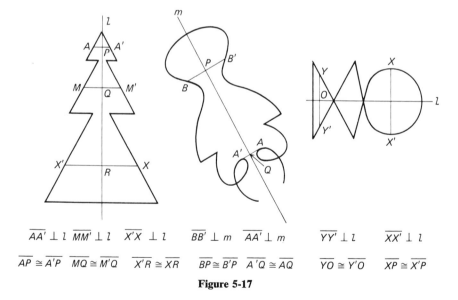

$\overline{AA'} \perp l$	$\overline{MM'} \perp l$	$\overline{X'X} \perp l$	$\overline{BB'} \perp m$	$\overline{AA'} \perp m$	$\overline{YY'} \perp l$	$\overline{XX'} \perp l$
$\overline{AP} \cong \overline{A'P}$	$\overline{MQ} \cong \overline{M'Q}$	$\overline{X'R} \cong \overline{XR}$	$\overline{BP} \cong \overline{B'P}$	$\overline{A'Q} \cong \overline{AQ}$	$\overline{YO} \cong \overline{Y'O}$	$\overline{XP} \cong \overline{X'P}$

Figure 5-17

The description of symmetry about a point is similar.

DEFINITION 5.18

A figure is **symmetric about a point** P, if for all points A of the figure there exists a symmetric partner A' for A (not necessarily distinct), so that $\overline{AA'}$ lies on P and such that $\overline{AP} \cong \overline{A'P}$.

This type of symmetry is shown in the parts of Fig. 5-18.

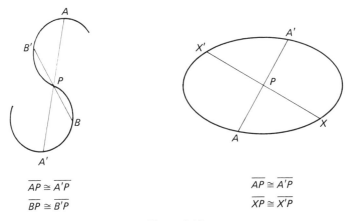

$$\overline{AP} \cong \overline{A'P}$$
$$\overline{BP} \cong \overline{B'P}$$

$$\overline{AP} \cong \overline{A'P}$$
$$\overline{XP} \cong \overline{X'P}$$

Figure 5-18

EXERCISE SET 5.5

1. Are regular polygons symmetric? If so, about a point? about a line?

2. If your answer to the last part of the above question is affirmative, tell how many axes of symmetry an equilateral triangle has? A square? A regular pentagon? A regular hexagon? A regular n-gon?

3. How many capital letters in the Latin alphabet are symmetric?

4. How many of the Arabic numerals are symmetric?

5. You perhaps have noticed from some of the examples and some of the exercises that it is possible for a figure to be symmetric about a line as well as symmetric about a point. Try to characterize or describe all such figures that possess both symmetries.

6. As noted, symmetric figures consist of two parts that are congruent to

each other. Therefore, the picture of any symmetric figure may be generated by tracing over the picture of any figure twice and then putting the two parts together. Make pictures of figures that are symmetric about a line and about a point by using the figures below as the initial pictures to be traced.

7. Designs that might appear on an article of clothing can often be generated by repeatedly reflecting (about a line of symmetry) and rotating (about a point of symmetry) some simple geometric figure. What is the simplest figure one can use to generate the following patterns?

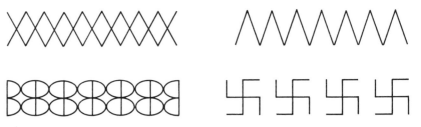

8. A popular task that elementary school teachers often give their students is the cutting out of "snowflakes." This task is done by folding over a rectangular sheet of paper vertically down the middle, and then folding it over again horizontally, and then perhaps a third time on an angle through the common point of the first two folds so that the two original

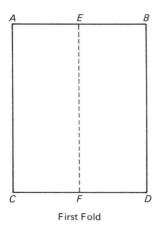

First Fold

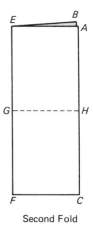

Second Fold

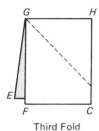

Third Fold

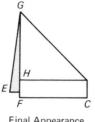

Final Appearance
Before Cutting

folded edges coincide, and then finally cutting various designs with
scissors. Is the final result, the "snowflake," a symmetric figure?

SUGGESTED READINGS

   [2]

   [3]

   [11] Ch. 3

   [13] Ch. 6

   [21] Ch. 6

   [22] Ch. 8

   [26] Ch. 7

6

SIMILARITY

In the last chapter, we discussed the idea of congruent figures—those that have the same shape and size. In this chapter, we shall look at the concept of *similar* figures—those that have the same shape but not necessarily the same size.

6.1 PROPORTIONALITIES

For polygons to be congruent, corresponding angles and corresponding segments must be congruent; for similar polygons the demand on the angles is the same, but the corresponding segments need not be congruent, only proportional. It is the purpose of this section to make clear the concept of a proportionality. The definition of segment congruence was made in terms of the measures of segments—real numbers. Real numbers are also pertinent for the discussion of a proportionality.

Consider two ordered sets of real numbers,

$$\{a_1, a_2, a_3, \ldots\} \quad \text{and} \quad \{b_1, b_2, b_3, \ldots\}$$

If it is true that

$$\frac{a_1}{b_1} = \frac{a_2}{b_2} = \frac{a_3}{b_3} = \cdots$$

then those sets are said to be **proportional** to each other, or we say a **proportionality** exists between the two sets. The constant number

$$k = \frac{a_1}{b_1} = \frac{a_2}{b_2} = \frac{a_3}{b_3} = \cdots$$

is called the **constant of proportionality.** This proportionality relationship on the class of all ordered sets of numbers is an equivalence relation.

When two sets consist of only two elements each, such as $\{a, b\}$ and $\{c, d\}$, the single equation resulting from a proportionality between those two sets is often called simply a **proportion.**

To speak of proportionalities between sets of segments is not quite proper, since the term applies correctly only to sets of numbers; however, to avoid excessive verbiage, the convention is made to use the term "proportional" with segments, even though it properly refers to the measure of those segments.

A few numerical examples are in order.

EXAMPLES

(a) $\{1, 3, 5, 6, 9\}$ and $\{3, 9, 15, 18, 27\}$ are proportional, since

$$\frac{1}{3} = \frac{3}{9} = \frac{5}{15} = \frac{6}{18} = \frac{9}{27}$$

(b) $\{2, 5, 7\}$ is proportional to $\{0.5, 1.25, 1.75\}$ since

$$\frac{2}{0.5} = \frac{5}{1.25} = \frac{7}{1.75} = 4$$

Because proportionalities can be expressed as algebraic equations involving ratios of real numbers, some useful theorems follow immediately. The proofs are called for in the exercises.

THEOREM 6.1

If $\dfrac{a}{b} = \dfrac{c}{d}$, then $\dfrac{a}{c} = \dfrac{b}{d}$.

THEOREM 6.2

If $\dfrac{a}{b} = \dfrac{c}{d}$, then $\dfrac{b}{a} = \dfrac{d}{c}$.

THEOREM 5.3

If $\dfrac{a}{b} = \dfrac{c}{d}$, then $\dfrac{c}{a} = \dfrac{d}{b}$.

THEOREM 6.4

If $\dfrac{a}{b} = \dfrac{c}{d}$, then $\dfrac{a+b}{b} = \dfrac{c+d}{d}$.

EXERCISE SET 6.1

1. Given that a proportionality exists between the following pairs of sets of numbers, supply the missing numbers.

   (a) $\{3, 4, ?, 9, 11, ?, 2\}$ and $\{?, 6, 8, 13.5, ?, 12, ?\}$

   (b) $\{\frac{1}{3}, \frac{2}{5}, ?, ?, \frac{7}{4}\}$ and $\{\frac{1}{6}, ?, \frac{2}{5}, 1\frac{1}{2}, ?\}$

   (c) $\{3, 5, 6.5, ?, ?\}$ and $\{?, 7.5, ?, 15, 4.2\}$

2. Find the value for x in the following proportions:

   (a) $\dfrac{x}{7} = \dfrac{2}{3.5}$ (b) $\dfrac{4}{9} = \dfrac{x}{8}$ (c) $\dfrac{2}{x} = \dfrac{x}{10}$ (d) $\dfrac{x}{4.5} = \dfrac{x-3}{9}$

3. Prove Theorem 6.1.

4. Prove Theorem 6.2.

5. Prove Theorem 6.3.   (*Hint*: Use Theorems 6.1 and 6.2.)

6. Prove Theorem 6.4.   (*Hint*: Add 1 to both sides of $\dfrac{a}{b} = \dfrac{c}{d}$ and then simplify.)

6.2   SIMILIAR TRIANGLES

Similar figures have the same shape but not necessarily the same size. A description can be given by saying that a picture of one figure is a scale drawing of the picture of the other. This description is not too good if three-dimensional figures are involved, but then we might say that one is a scale model of the other.   These descriptions help to convey the idea of similarity, but they lack the desirable mathematical quality of being precise. This precision is gained by making use of the triangle, which is the simplest of all polygons.   Plane figures other than polygons will be discussed later in the chapter.

The one-to-one correspondence is the keynote.   The notation for correspondences between triangles shall be the same as it was for triangle congruence. The correspondence $ABC \leftrightarrow DEF$ still establishes the six pairs of corresponding parts, the three pairs of angles and the three pairs of sides. With the concept of a one-to-one correspondence and the idea of proportionality developed in the previous section, we make the following definition of similar triangles.

DEFINITION 6.1

Two triangles, not necessarily distinct, are **similar** if there exists a one-to-one correspondence between vertices of the two triangles such that the corresponding angles are congruent and the corresponding sides are proportional.

The symbol used for similar triangles is "$\sim$."   For example, we would write: $\triangle ABC \sim \triangle DEF$.   The usage of this symbolism will automatically convey the following information:

$$\angle ABC \cong \angle DEF, \quad \angle BCA \cong \angle EFD, \quad \angle CAB \cong \angle FDE, \quad \text{and}$$

$$\frac{\overline{AB}}{\overline{DE}} = \frac{\overline{AC}}{\overline{DF}} = \frac{\overline{BC}}{\overline{EF}}$$

where the equality between the ratios of the sides really stands for the measures of those sides, as agreed on previously.

As with so many of the other relationships that have been studied, this is an equivalence relation defined on the set of all triangles.   This fact needs a bit of amplification, because an important result is embedded in it. If similarity is to be an equivalence relation, then it must have the property of being reflexive; that is, a triangle must be similar to itself.   But not only does a triangle have the same shape as itself, it also has the same size, which is equivalent to saying that it is congruent to itself.   Thus, triangle congruence is a special case of triangle similarity.   In triangle congruence, the sides are proportional to each other—the constant of proportionality happens to be the number "1."

This fact can be generalized by saying that congruence of any geometric figures is a special case of similarity of geometric figures.

It is of interest to ask what minimum conditions are necessary to ensure that a one-to-one correspondence leads to similar triangles. The statements given for congruence were made as postulates—that is, without proof; the ones for similarity may be proven as theorems. In order to prove these theorems, certain other theorems are needed. These happen to be theorems relating to parallel lines, which have not yet been discussed in this text.

Since similarity and congruence are so closely related, it has been judged to be pedagogically more sound to proceed by introducing similarity before parallelism. From the rigorous mathematical point of view regarding logical development of subject, this choice is not the best. As a result, the statements on similarity conditions will be labeled as postulates rather than theorems. The student is strongly urged to consult any of the many fine textbooks on geometry at the high school or college level to see how the following statements can be proven by means of previously established results on parallel lines.

POSTULATE 6.5 (AAA Postulate for Similarity)

If there exists a one-to-one correspondence between the vertices of two triangles such that the corresponding angles are congruent, then the two triangles are similar to each other.

This postulate is illustrated by several examples shown in Fig. 6-1.

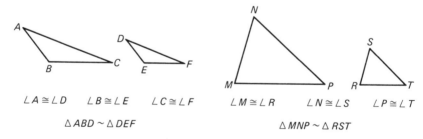

$\angle A \cong \angle D$ $\angle B \cong \angle E$ $\angle C \cong \angle F$ $\angle M \cong \angle R$ $\angle N \cong \angle S$ $\angle P \cong \angle T$

$\triangle ABD \sim \triangle DEF$ $\triangle MNP \sim \triangle RST$

Figure 6-1

The similar triangles are placed in the same relative positions to make it easier to notice the similarity of the shapes of the triangles.

It was mentioned earlier that the postulates of this section could have been proven as theorems if we were armed with some results following from relationships about parallel lines.   There is one particular result of the study of parallels that will be mentioned here.   It is the theorem that states that the sum of the measures of the angles of any triangle is 180.   Using this theorem, we shall state a theorem whose proof should be obvious.

THEOREM 6.6 (Third Angle Theorem)

If two angles of one triangle are congruent respectively to two angles of a second triangle, then the third angles of each are congruent to each other.

Now using Theorem 6.6 and the AAA Similarity Postulate, we derive the following theorem.

THEOREM 6.7 (AA Similarity Theorem)

If there exists a one-to-one correspondence between the vertices of two triangles such that two angles of one are congruent respectively to the corresponding angles of the second, then the triangles are similar.

POSTULATE 6.8 (SSS Postulate for Similarity)

If there exists a one-to-one correspondence between the vertices of two triangles such that the corresponding sides are proportional, then the triangles are similar.

Several examples are pictured in Fig. 6-2.

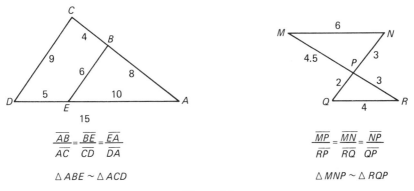

$$\frac{\overline{AB}}{\overline{AC}} = \frac{\overline{BE}}{\overline{CD}} = \frac{\overline{EA}}{\overline{DA}}$$

$$\triangle ABE \sim \triangle ACD$$

$$\frac{\overline{MP}}{\overline{RP}} = \frac{\overline{MN}}{\overline{RQ}} = \frac{\overline{NP}}{\overline{QP}}$$

$$\triangle MNP \sim \triangle RQP$$

Figure 6-2

POSTULATE 6.9 (SAS Postulate for Similarity)

If there exists a one-to-one correspondence between the vertices of two triangles such that two corresponding sides are proportional and their included angles are congruent, then the triangles are similar.

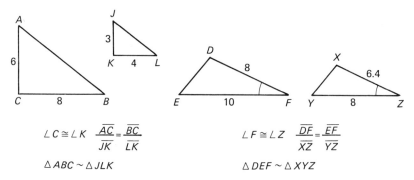

$$\angle C \cong \angle K \quad \frac{AC}{JK} = \frac{BC}{LK}$$

$$\triangle ABC \sim \triangle JLK$$

$$\angle F \cong \angle Z \quad \frac{DF}{XZ} = \frac{EF}{YZ}$$

$$\triangle DEF \sim \triangle XYZ$$

Figure 6-3

Examples of this postulate are shown in Fig. 6-3.

6.3   SIMILAR FIGURES

Similar polygons can be handled easily with the same techniques we used with congruent polygons. We state the following definition.

DEFINITION 6.2

Two polygons are **similar** if there exists a one-to-one correspondence between their vertices such that corresponding angles are congruent and the corresponding sides are proportional.

Notice that this definition is essentially identical to Definition 6.1, which defined similar triangles. This likeness is to be expected, since a triangle is a special case of a polygon.

The AAA Postulate for Similar Triangles states that congruence of corresponding angles ensures the existence of a similarity between triangles. The fact that such a postulate does not hold for polygons can be demonstrated by looking at a counterexample involving a square and a rectangle, which are pictured in Fig. 6-4. Both the rectangle and the square have four right angles; thus, we see that corresponding angles are congruent. Yet, the sides are not proportional: $m(\overline{AB}) = 2m(\overline{RS})$, whereas $m(\overline{ST}) = 2m(\overline{BC})$. For two polygons to be similar, all sides of one polygon must have a measure that is a constant times the measure of the corresponding sides of the second, provided the angles are congruent.

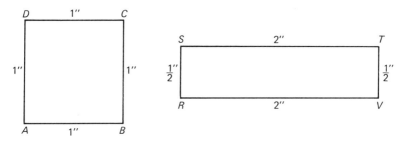

Figure 6-4

It can also be shown that proportionality of respective sides is not sufficient to ensure that two polygons are similar. An example will be used with a square and a rhombus. (A rhombus is a four-sided polygon with all four sides having the same measure.) A square is a rhombus, but a rhombus need not be a square. Figure 6-5 illustrates the situation with

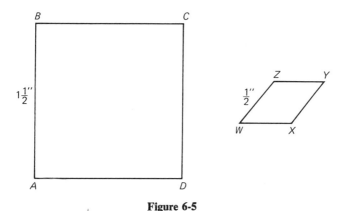

Figure 6-5

a square and a rhombus, where the constant of proportionality between the sides is 3. But the figure shows that the two polygons do not have the same shape.

Polygons are easy to describe because of their sides and angles—but how about some curve that has no angles or sides? As with triangles and polygons, we demand the possibility of establishing a one-to-one correspondence between the sets of points that are the figures in question. Consider any two distinct points of one of the figures and the corresponding points of the other figure. Between the two points of one of the figures there exists a segment, the points of which need not be points of the figure. In like manner, there exists a segment determined by the two points of the second figure. Call these segments **corresponding segments.** For all possible pairs of points of one figure and the corresponding pairs of the second figure, there exist pairs of corresponding segments. If all these pairs of segments are proportional, then we say the figures are similar. Figure 6-6

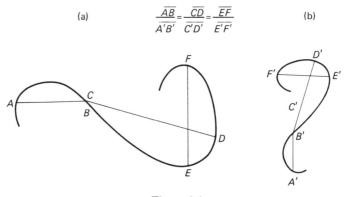

$$\frac{\overline{AB}}{A'B'} = \frac{\overline{CD}}{C'D'} = \frac{\overline{EF}}{E'F'}$$

(a)   (b)

Figure 6-6

illustrates two similar figures, where the like-named points are those that correspond to each other. The three like-named segments are but a few of the many pairs of segments that do exist. The figure pictured in part (a) is similar to the figure pictured in part (b).

EXERCISE SET 6.2

1. In each of the following parts shown on pages 123 and 124, there are pictured a pair of triangles that can be proven similar. Indicate the correspondence between vertices that ensures that they are similar and the statement on similarity of triangles (AA, SSS, or SAS) that proves that they are similar.

(a)

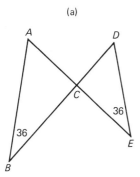

(b)

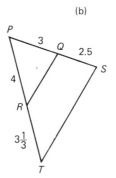

(c)

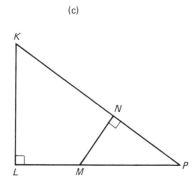

(d)

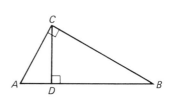

(e)

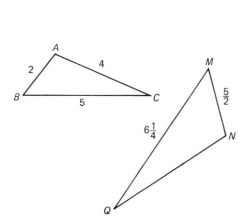

(f)

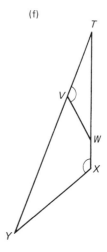

(g)   (h)

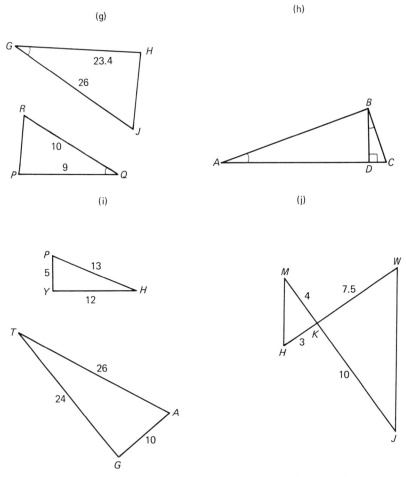

(i)   (j)

2. Give the reasons for the statements in the following proof.

Theorem: If two triangles are similar, then altitudes from corresponding vertices are proportional to corresponding sides of the triangles.

Proof: Refer to the figure below.   We hypothesize $\triangle ABC \sim \triangle MNQ$, using the correspondence $ABC \leftrightarrow MNQ$. $\overline{BD}$ is the altitude to $\overline{AC}$; $\overline{NP}$ is the altitude to $\overline{MQ}$.

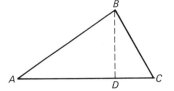   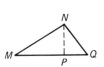

Statement	Reason
1. $\triangle ABC \sim \triangle MNQ$	1. ?
2. $\angle BAD \cong \angle NMP$	2. Definition of triangle similarity
3. $\overline{BD}$ is altitude to $\overline{AC}$. $\overline{NP}$ is altitude to $\overline{MQ}$.	3. ?
4. $\angle BDA$ is a right angle. $\angle NPM$ is a right angle.	4. ?
5. $\angle BDA \cong \angle NPM$	5. ?
6. Therefore, $\triangle BAD \sim \triangle NMP$	6. ? (Using the correspondence $BAD \leftrightarrow NMP$)
7. $\overline{BD}/\overline{NP} = \overline{BA}/\overline{NM}$	7. Definition of triangle similarity ($\triangle BAD \sim \triangle NMP$)
8. $\overline{BA}/\overline{NM} = \overline{AC}/\overline{MQ} = \overline{CB}/\overline{QN}$	8. Definition of triangle similarity ($\triangle ? \sim \triangle ?$)
9. $\overline{BD}/\overline{NP} = \overline{BA}/\overline{NM} = \overline{AC}/\overline{MQ} = \overline{CB}/\overline{QN}$	9. ?

3. Problem 2 of this set proved a theorem regarding corresponding altitudes in similar triangles. Is the theorem true if the word "altitude" is replaced by the word "median" or by the words "angle bisector"? If so, why? If not, why not?

6.4 THE PYTHAGOREAN THEOREM

One of the most famous theorems of all mathematics is the Pythagorean Theorem, which deals with the measures of the sides of a right triangle. Named after the Greek mathematician and philosopher, Pythagoras, who lived several centuries before Christ, this particular theorem has proved to be of great interest to mathematicians of all types, professional as well as amateur. Several hundred different proofs of it are known, one of which was discovered by James Garfield, a former president of the United States. The proof given in this section depends on a preliminary result that will be shown first. The preliminary theorem uses similar triangles in its proof.

THEOREM 6.10

The altitude to the hypotenuse of a right triangle forms two new triangles, each of which is similar to the other and to the original right triangle.

Proof: Refer to Fig. 6-7, where $\overline{BD}$ is the altitude to hypotenuse $\overline{AC}$. The first correspondence to consider is $ABC \leftrightarrow ADB$; the second is $ABC \leftrightarrow BDC$.

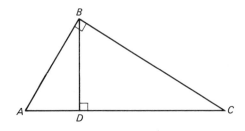

Figure 6-7

Statement	Reason
1. $\angle A \cong \angle A$	1. Reflexive property of angle congruence
2. $\angle ABC \cong \angle ADB$	2. They are both right angles.
3. $\triangle ABC \sim \triangle ADB$	3. Theorem 6.6 (the AA theorem with the correspondence $ABC \leftrightarrow ADB$)
4. $\angle C \cong \angle C$	4. Same as Reason 1
5. $\angle ABC \cong \angle BDC$	5. Same as Reason 2
6. $\triangle ABC \sim \triangle BDC$	6. Same as Reason 3, except using the correspondence $ABC \leftrightarrow BDC$
7. $\triangle ABC \sim \triangle ADB \sim \triangle BDC$	7. Transitivity of triangle similarity

The reason we proved this result is that we wish to use the property of similar triangles, which says corresponding sides are proportional. These proportions are used in the proof of the Pythagorean Theorem.

Before the theorem is formally stated, some notation needs to be introduced. If we refer to Fig. 6-8, lowercase Latin letters placed alongside

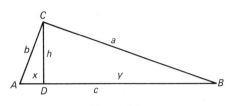

Figure 6-8

the various segments are used to represent the measure of those segments. For example, $b = m(\overline{AC})$ and $h = m(\overline{CD})$. Wherever possible, a lowercase letter corresponds to a side opposite an angle named by the same letter in upper case. For example, c is the measure of the side opposite $\angle C$.

One other thing needs to be discussed before the proof is given. In the proof, we shall need to make use of the fact that, as shown in Fig. 6-8, $x + y = c$, or putting it in terms of segments, $\overline{AD} + \overline{DB} = \overline{AB}$. Addition of real numbers, such as $x + y$, has meaning, but addition of segments, which are sets, does not. To be logically correct, a statement must be made that gives meaning to the addition of segments. If segment addition is to be similar to ordinary addition, we must demand closure—that is, the result of adding two segments must be a segment. This statement is given in the following postulate.

POSTULATE 6.11 (Segment Addition Postulate)

If A, B, and C are collinear points such that B is between A and C, then it is possible to "add" $\overline{AB}$ and $\overline{BC}$ to get the "sum" $\overline{AC}$, provided $m(\overline{AB}) + m(\overline{BC}) = m(\overline{AC})$.

THEOREM 6.12 (The Pythagorean Theorem)

In a right triangle, the square of the measure of the hypotenuse is equal to the sum of the squares of the measures of the other two sides.

Proof: Refer to Fig. 6-8. The aim is to show $a^2 + b^2 = c^2$. Using Theorem 6-10 and noting that $\overline{CD}$ is the altitude to hypotenuse AB, we derive the similarity:

$$\triangle ADC \sim \triangle ACB$$

This leads to:

$$\frac{\overline{AD}}{\overline{AC}} = \frac{\overline{AC}}{\overline{AB}}$$

which, in terms of measures, is:

$$\frac{x}{b} = \frac{b}{c}$$

Using algebra, we get:

$$x = \frac{b^2}{c} \tag{6-1}$$

In a similar fashion, using Theorem 6.10, we derive another similarity:

$$\triangle DCB \sim \triangle CAB$$

This leads to:

$$\frac{\overline{DB}}{\overline{CB}} = \frac{\overline{CB}}{\overline{AB}}$$

which, in terms of measures, is:

$$\frac{y}{a} = \frac{a}{c}$$

Using algebra, we get:

$$y = \frac{a^2}{c} \tag{6-2}$$

Combining Eqs. (6-1) and (6-2) gives:

$$x + y = \frac{b^2}{c} + \frac{a^2}{c} = \frac{b^2 + a^2}{c} \tag{6-3}$$

Recalling that $x + y = c$, we may substitute this into Eq. (6-3) to get:

$$c = \frac{b^2 + a^2}{c}$$

which simplifies into the desired result:

$$c^2 = a^2 + b^2$$

The converse of the Pythagorean Theorem is also true, although it will not be proven here.

THEOREM 6.13 (Converse of the Pythagorean Theorem)

If $\triangle ABC$ has sides with measures a, b, and c, respectively, such that $a^2 + b^2 = c^2$, then $\triangle ABC$ is a right triangle with the right angle at C.

Triples of positive integers that satisfy the Pythagorean Theorem are called **Pythagorean triples**. Some of the well-known ones are:

3, 4, 5,   since $3^2 + 4^2 = 5^2$

5, 12, 13,   since $5^2 + 12^2 = 13^2$

8, 15, 17,   since $8^2 + 15^2 = 17^2$

Any integral multiples of Pythagorean triples are also Pythagorean triples. For example, 6, 8, 10 and 15, 36, 39 are also Pythagorean triples. Formulas are known that produce such triples, although they will not be presented in this text.

6.5 PRACTICAL APPLICATIONS OF SIMILAR FIGURES

In an earlier section, similar figures were intuitively described as being scale drawings or scale models of each other. Now that similar figures have been more precisely defined, we can "look at the other side of the coin" to see how knowledge of similar figures leads to an understanding of scale drawings. Two examples of scale drawings are plans, or blueprints, of a building and maps. The map or plan that we see is the picture of the smaller of two figures, whereas the larger figure has a size too prohibitive to picture it in actual size on a piece of paper.

A line segment connecting two points pictured on a map has a measure that, when divided by the measure of the segment connecting the two real points represented by the points of the map, gives the constant of proportionality that exists as a constant for all such segments and their real counterparts. This constant of proportionality is often given on the map by the word "*scale*," such as "Scale: 1 inch: 20 miles." Sometimes, a picture of a ruler is given, as shown in Fig. 6-9. Thus, by using the scale, a person can find the real distance between two points by finding the distance between the points of the map that represent these points and then multiplying this distance by the constant of proportionality. Often the distance can be

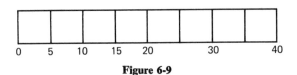

Figure 6-9

WAY COUNTY

By   Hi   Free

Thru   Toll

Scale: $\frac{1}{4}$ inch: 5 miles

Figure 6-10

found by representing the scale as a proportion and then solving the proportion by algebra. An illustration is given in Fig. 6-10.

EXAMPLES

(a) From Thru to Toll. On the map, the distance is approximately $\frac{9}{4}$ in.

$$\frac{\frac{1}{4}}{5} = \frac{\frac{9}{4}}{x}$$

$$\frac{1}{4}x = \frac{45}{4}$$

$$x = 45$$

Thus, the distance from Thru to Toll is approximately 45 miles.

(b) From By to Hi. On the map, the distance is approximately $\frac{7}{4}$ inches.

$$\frac{\frac{1}{4}}{5} = \frac{\frac{7}{4}}{x}$$

$$\frac{1}{4}x = \frac{35}{4}$$

$$x = 35$$

Thus, the distance from By to Hi is approximately 35 miles.

Another valuable use of similar figures, and especially similar triangles, is in indirect measurement. When a measuring instrument can be placed directly adjoining the figure to be measured, the measurement is said to be **direct**. When a measurement is derived by some other means whereby the measuring instrument is not used directly, then the measurement is said to be **indirect**. Examples might be finding the height of a tree without climbing to its top, finding the distance across a lake without actually going across it, finding the distance to a shoreline from a boat without actually going to the shore. Several examples will be given to show how this can be done.

EXAMPLE

(c) To find the height of a tree without climbing it.

In Fig. 6-11, the dotted lines indicate the rays of the sun. Since the sun is so far from the earth, it is reasonable to assume that the rays of the sun strike the earth in parallel lines. Thus, the angles at A and B are congruent. If the small picture at the

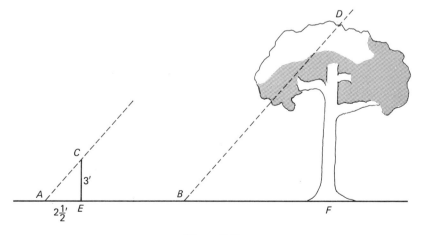

Figure 6-11

left represents a pole 3 ft high making right angles with the ground and if we also asume that the tree is at right angles with the ground, then $\triangle ACE$ is similar to $\triangle BDF$ by the AA theorem. The sides are then proportional; this fact leads to the proportion

$\dfrac{\overline{AE}}{\overline{BF}} = \dfrac{\overline{CE}}{\overline{DF}}$, which, in terms of the hypothesized situation, may be

stated as: shadow is to shadow as height is to height. The height of the pole is approximately 3 ft, the length of its shadow approximately $2\frac{1}{2}$ ft, and the length of the shadow of the tree is approximately 40 ft.

The proportion to be solved is:

$$\frac{2\frac{1}{2}}{40} = \frac{3}{x}$$

$$2\frac{1}{2}x = 120$$

$$x = 48$$

Therefore, the tree is approximately 48 ft tall.

EXAMPLE

(d) Finding the distance across a lake without crossing it.

In Fig. 6-12, $\overline{AB}$ represents the distance across the lake.

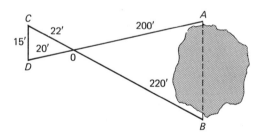

Figure 6-12

By choosing some convenient point O (it may be a bush, or a rock, or a stake put in the ground), the distances $\overline{AO}$ and $\overline{BO}$ are found by direct measurement. Then points C and D are located in a special way. $\overline{CB}$ and $\overline{DA}$ must be straight line segments, and $\overline{CO}$ and $\overline{DO}$ must have lengths that are the same constant times the lengths of $\overline{BO}$ and $\overline{AO}$, respectively. In this example, they are each one-tenth of the length. Therefore, $\triangle AOB \sim \triangle DOC$ by the SAS Similarity Postulate. (Why is $\angle AOB \cong \angle DOC$?)

Since the triangles are similar, all corresponding sides are proportional; $\overline{CD}$ corresponds to $\overline{BA}$, and $\overline{CD}$ is found by measurement to be approximately 15 ft long. Thus, the lake is approximately 150 feet across from point A to point B.

EXAMPLE

(e) Finding the distance to shore from a boat without actually going to shore.

In Fig. 6-13, a steamship is moving along a straight course determined by the points B and C in the picture. The captain

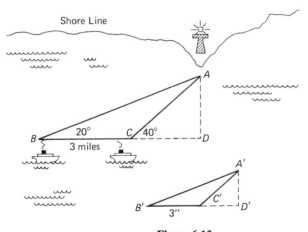

Figure 6-13

wishes to know how far his ship will be from shore when it passes directly opposite the lighthouse at Promontory Point, which is at point A of the picture. We assume that he has an instrument that can measure the size of an angle between the front (bow) of his ship and a point on the shoreline. Suppose at a given moment he measures the angle and finds it to be 20° at point B. If his ship is traveling through the water at the rate of 9 mph, he can wait 20 minutes and take another measurement of the angle between bow and the lighthouse. He knows he has traveled a distance of 3 miles, which is shown in the top figure as the length of $\overline{BC}$. If he finds the angle to be 40° this time, he then has all the information he needs to get an approximation of his distance from A when he is at point D.

This approximation can be found by making a scale drawing of the top part of the figure shown above. The scale drawing has been represented by $\triangle A'B'C'$, which is made similar to $\triangle ABC$ by making $\angle B = \angle B'$ and $\angle ACB = \angle A'C'B'$. $\overline{AD}$ and $\overline{A'D'}$ are the respective altitudes of the two triangles and are, therefore, in the same proportion as the sides of the two triangles. (Recall Exercise 2 of Exercise Set 6.2.) The length of $\overline{B'C'}$ is really immaterial, but it has been shown as 3 in. to make the mathematics easy. The captain measures $\overline{A'D'}$ and finds it to be approximately $2\frac{1}{4}$ in. long. Thus, he concludes that he will be approximately $2\frac{1}{4}$ miles offshore from Promontory Point Lighthouse when he passes directly opposite it.

EXERCISE SET 6.3

1. In each of the following right triangles pictured, use the Pythagorean Theorem to find the length of the sides whose measures are indicated by x.

(a)

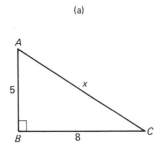

(b)

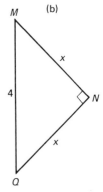

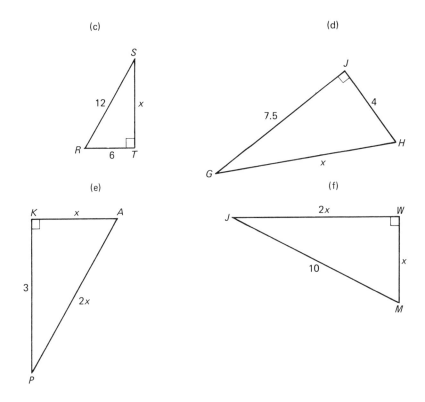

(c)

(d)

(e)

(f)

2. Refer back to Fig. 6-11. Find the height of the tree if:

(a) $\overline{AE} = 5$ ft $\overline{CE} = 2$ ft 6 in. $\overline{BF} = 60$ ft

(b) $\overline{AE} = 6\frac{2}{3}$ ft $\overline{CE} = 4$ ft $\overline{BF} = 48$ ft

(c) $\overline{AE} = 8$ ft 4 in. $\overline{CE} = 4$ ft 9 in. $\overline{BF} = 50$ ft

(d) $\overline{AE} = 4$ ft $\overline{CE} = 6.4$ ft $\overline{BF} = 39.6$ ft

3. Refer back to Fig. 6-12. Find the distance across the lake if:

(a) $\overline{AO} = 150$ ft $\overline{DO} = 7.5$ ft $\overline{CD} = 21$ ft

(b) $\overline{AO} = 180$ ft $\overline{DO} = 12$ ft $\overline{CD} = 16$ ft

(c) $\overline{BO} = 196$ ft $\overline{CO} = 20.4$ ft $\overline{CD} = 12.5$ ft

(d) $\overline{BO} = 432$ ft $\overline{CO} = 24$ ft $\overline{CD} = 12$ ft

4. In the figure shown on page 136, a method for finding the distance across a river is shown. By using an angle measuring device, we have constructed right angles at points A and D. By making $\overline{BE}$ a straight line

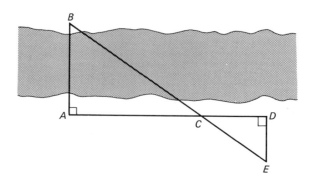

segment through point C, we get similar triangles $\triangle ABC$ and $\triangle DEC$.
How? $\overline{AC}$, $\overline{CD}$, and $\overline{DE}$ can be found by direct measurement. Find
$\overline{AB}$ if:

(a) $\overline{AC} = 12$ ft $\overline{CD} = 4$ ft $\overline{DE} = 3$ ft

(b) $\overline{AC} = 25$ ft $\overline{CD} = 10$ ft $\overline{DE} = 4$ ft

(c) $\overline{AC} = 36$ ft $\overline{CD} = 4.5$ ft $\overline{DE} = 2$ ft 6 in.

(d) $\overline{AC} = 40$ ft $\overline{CD} = 5.6$ ft $\overline{DE} = 3$ ft $4\frac{1}{2}$ in.

5. Formulate an Angle-Addition Postulate similar to the Segment-Addition
Postulate of Section 6.4. Be sure that conditions are stipulated, so that
the sum of two angles wil be an angle.

6.6   NUMERICAL TRIGONOMETRY

Trigonometry is a branch of mathematics that has proved to be of interest
to man for a long, long time. It has been a useful tool to persons in many
diverse trades; much of the interest can be traced to its practicality. Today,
students of the subject value it for its own intrinsic interest as an example
of a mathematical system. There are scores of excellent textbooks written
on the subject, and it is strongly recommended that the interested reader
study any one of these.

The present section of this text deals with but a small portion of the
subject, namely numerical trigonometry of the right triangle. Numerical
trigonometry is actually a special instance of indirect measurement. The
development of the topic depends upon the theory of similar triangles and
makes particular use of the AA Similarity Theorem.

In the first example and all other examples and discussions that follow, we shall be talking exclusively about right triangles, unless otherwise mentioned. In Fig. 6-14, there are pictured two right triangles, each with an

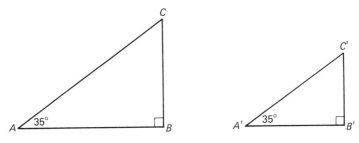

Figure 6-14

acute angle of $35°$. Because all right angles are congruent to each other and because two angles with the same measure are congruent, the two triangles are similar to each other by the AA theorem. It is, in fact, true that *all* right triangles with a $35°$ angle as one of the acute angles are similar to each other. Being similar, their corresponding sides are proportional. For the triangles of Fig. 6-14, we have

$$\frac{\overline{BC}}{\overline{B'C'}} = \frac{\overline{AC}}{\overline{A'C'}}$$

which by the use of Theorem 6.1 can be written in the form

$$\frac{\overline{BC}}{\overline{AC}} = \frac{\overline{B'C'}}{\overline{A'C'}}$$

The latter equation is noteworthy, since both the left and right sides of the equation involve ratios of sides in a *single* triangle. This idea of the ratio involving sides of a single triangle is the heart of the study of numerical trigonometry.

In the investigation of this topic, it is most important to remember that the above ratios and proportions, while being written in terms of segments, are really ratios and proportions dealing with the measures of those segments and, as such, are ratios and proportions of real numbers.

Each of the sides of the triangle has a particular name, given with reference to an acute angle of the triangle. The names are shown in Fig. 6-15. If we consider $\angle A$ as the acute angle, $\overline{BC}$ is called the side **opposite**

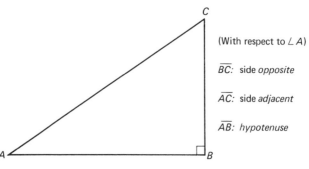

Figure 6-15

$\angle A$; $\overline{AC}$ is called the side **adjacent** to $\angle A$; $\overline{BA}$ is called by its accustomed name, the **hypotenuse.**

A ratio is often thought of as the quotient of two numbers; thus, with three distinct sides of a triangle, there are six separate ratios that can be formed:

$$\frac{\overline{BC}}{\overline{AB}}, \quad \frac{\overline{AC}}{\overline{AB}}, \quad \frac{\overline{BC}}{\overline{AC}}, \quad \frac{\overline{AB}}{\overline{BC}}, \quad \frac{\overline{AB}}{\overline{AC}}, \quad \frac{\overline{AC}}{\overline{BC}}$$

Since the last three are merely the reciprocals (multiplicative inverses) of the first three, they are not given the attention accorded the first three. The first three involve the three possible pairings that can be made from the three distinct sides of a right triangle. All of these ratios have their own name by which they are remembered; together they are called the six **trigonometric ratios.**

In a right triangle:

DEFINITION 6.3

The ratio of the side opposite an acute angle divided by the hypotenuse is called the **sine** of the angle, and is written: sin A. Thus:

$$\sin A = \frac{\text{opp.}}{\text{hyp.}}$$

DEFINITION 6.4

The ratio of the side adjacent to an acute angle divided by the hypotenuse is called the **cosine** of the angle, and is written: cos A. Thus:

$$\cos A = \frac{\text{adj.}}{\text{hyp.}}$$

DEFINITION 6.5

The ratio of the side opposite an acute angle divided by the side adjacent to the angle is called the **tangent** of the angle, and is written: tan A. Thus:

$$\tan A = \frac{\text{opp.}}{\text{adj.}}$$

DEFINITION 6.6

The ratio that is the reciprocal of the tangent of an angle is called the **cotangent** of the angle, and is written: cot A. Thus:

$$\cot A = \frac{\text{adj.}}{\text{opp.}}$$

DEFINITION 6.7

The ratio that is the reciprocal of the cosine of an angle is called the **secant** of the angle, and is written: sec A. Thus:

$$\sec A = \frac{\text{hyp.}}{\text{adj.}}$$

DEFINITION 6.8

The ratio that is the reciprocal of the sine of an angle is called the cosecant of the angle, and is written: csc A. Thus:

$$\csc A = \frac{\text{hyp.}}{\text{opp.}}$$

The important thing to remember is that the values of these ratios depend only on (or are a function of) the measure of the acute angle.

The theory of similar triangles, as exemplified by the discussion following Fig. 6-14, assures us of this fact. Therefore, if you were to draw any-sized right triangle with an acute angle of 35°, the value of the ratio of the side opposite that angle divided by the hypotenuse of that triangle (that is, the sine of 35°) would be a fixed number. This fixed number is the same as it was thousands of years ago—as it is now—as it will be thousands of years from now—it will not change. This value, as well as any other value of a sine, cosine, or tangent of any acute angle can be found in a table of trigonometric ratios. Such a table has been reproduced in this text and appears as Table 6-1 on facing page 141.

What is the import of all this discussion? In essence, this theory allows the "solution" any right triangle, given the measure of one of its sides and one of the acute angles or given the measure of two of its sides. It enables a person to find indirect measurements of a right triangle without resorting to a second triangle, which is similar to it. Actually, the second triangle is incorporated into the table of trigonometric ratios. This theory is best explained by resorting to specific examples.

EXAMPLE

(a) In the right triangle pictured, $m(\angle A) = 25$, $a = 18$ ft. How long is $\overline{AB}$?

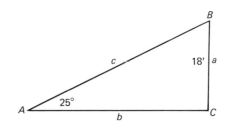

Solution:

$$\sin A = \frac{\text{opp.}}{\text{hyp.}} = \frac{a}{c}$$

$\sin 25° = 0.4226$ from Table 6-1. Thus,

$$\frac{a}{c} = 0.4226 \quad \text{or} \quad \frac{18}{c} = 0.4226$$

TABLE 6-1 Table of Trigonometric Ratios

Angle	sin	cos	tan	Angle	sin	cos	tan
				45°	.7071	.7071	1.000
1°	.0175	.9988	.0175	46°	.7193	.6947	1.036
2°	.0349	.9994	.0349	47°	.7314	.6820	1.072
3°	.0523	.9986	.0524	48°	.7431	.6691	1.111
4°	.0698	.9976	.0699	49°	.7547	.6561	1.150
5°	.0872	.9962	.0875	50°	.7660	.6428	1.192
6°	.1045	.9945	.1051	51°	.7771	.6293	1.235
7°	.1219	.9925	.1228	52°	.7880	.6157	1.280
8°	.1392	.9903	.1405	53°	.7986	.6018	1.327
9°	.1564	.9877	.1584	54°	.8090	.5878	1.376
10°	.1736	.9848	.1763	55°	.8912	.5736	1.428
11°	.1908	.9816	.1944	56°	.8290	.5592	1.483
12°	.2079	.9781	.2126	57°	.8387	.5446	1.540
13°	.2250	.9744	.2309	58°	.8480	.5299	1.600
14°	.2419	.9703	.2493	59°	.8572	.5150	1.664
15°	.2588	.9659	.2679	60°	.8660	.5000	1.732
16°	.2756	.9613	.2867	61°	.8746	.4848	1.804
17°	.2924	.9563	.3057	62°	.8829	.4695	1.881
18°	.3090	.9511	.3249	63°	.8910	.4540	1.963
19°	.3256	.9455	.3443	64°	.8988	.4384	2.050
20°	.3420	.9397	.3640	65°	.9063	.4226	2.145
21°	.3584	.9336	.3839	66°	.9135	.4067	2.246
22°	.3746	.9272	.4040	67°	.9205	.3907	2.356
23°	.3907	.9205	.4245	68°	.9272	.3746	2.475
24°	.4067	.9135	.4452	69°	.9336	.3584	2.605
25°	.4226	.9063	.4663	70°	.9397	.3420	2.747
26°	.4384	.8988	.4877	71°	.9455	.3256	2.904
27°	.4540	.8910	.5095	72°	.9511	.3090	3.078
28°	.4695	.8829	.5317	73°	.9563	.2924	3.271
29°	.4848	.8746	.5543	74°	.9613	.2756	3.487
30°	.5000	.8660	.5774	75°	.9659	.2588	3.732
31°	.5150	.8572	.6009	76°	.9703	.2419	4.011
32°	.5299	.8480	.6249	77°	.9744	.2250	4.331
33°	.5446	.8387	.6494	78°	.9781	.2079	4.705
34°	.5592	.8290	.6745	79°	.9816	.1908	5.145
35°	.5736	.8192	.7002	80°	.9848	.1736	5.671
36°	.5878	.8090	.7265	81°	.9877	.1564	6.314
37°	.6018	.7986	.7536	82°	.9903	.1392	7.115
38°	.6157	.7880	.7813	83°	.9925	.1219	8.144
39°	.6293	.7771	.8098	84°	.9945	.1045	9.514
40°	.6248	.7660	.8391	85°	.9962	.0872	11.43
41°	.6561	.7547.	8693	86°	.9976	.0698	14.30
42°	.6691	.7431	.9004	87°	.9986	.0523	19.08
43°	.6820	.7314	.9325	88°	.9994	.0349	28.64
44°	.6947	.7193	.9657	89°	.9998	.0175	57.29

Solving for c, we get $c = 18/0.4226 = 42.59$ ft.   Therefore, $\overline{AB}$ is approximately 42.6 ft long.

EXAMPLE

(b) In the picture below, $\overline{AC}$, the shadow of the tree, is 42 ft; the **angle of elevation,** that is, the angle at A between the horizontal and the line of sight to the top of the tree, is 40°.   How tall is the tree?

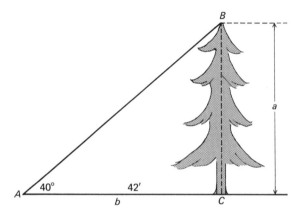

Solution:

$$\tan A = \frac{\text{opp.}}{\text{adj.}} = \frac{a}{b}$$

$\tan 40° = 0.8391$ from Table 6-1.   Thus,

$$\frac{a}{b} = 0.8391 \quad \text{or} \quad \frac{a}{42} = 0.8391$$

Solving for a, we get

$$a = (42)\,(0.8391) = 35.2422$$

Therefore, the tree is approximately 35.2 ft high.

EXAMPLE

(c) In the picture on page 143, $m(\angle B) = 60°$ and $\overline{BC} = 150$ ft.

Find the distance across the lake from point A to point B.

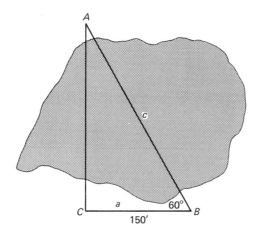

Solution:

$$\cos B = \frac{\text{adj.}}{\text{hyp.}} = \frac{m(\overline{BC})}{m(\overline{AB})} = \frac{a}{c}$$

$\cos 60° = 0.5000$ from Table 6-1. Thus,

$$\frac{a}{c} = 0.5000 \text{ or } \frac{150}{c} = 0.5000$$

Solving for c, we get

$$c = \frac{150}{0.5000} = 300$$

Therefore, the distance across the lake from A to B is approximately 300 ft.

EXAMPLE

(d) In a right triangle ABC with the right angle at C, $a = 3$ in. and $b = 4$ in. What is the size of $\angle A$, to the nearest degree?

Solution:

$$\tan A = \frac{a}{b} = \frac{3}{4} = 0.7500$$

If we look in Table 6-1 in the column headed "tan," the number closest to 0.7500 is 0.7536, which is the tangent of 37°. Thus, we conclude that $\angle A$ is approximately 37°.

In summary, if you wish to find the length of a missing side of a right triangle when given the length of one of the other sides and the measure of one of the acute angles, follow these steps:

1. Decide which of the three tabled trigonometric ratios (sine, cosine, or tangent) involves the two sides in question.
2. Write the appropriate ratio.
3. Set the ratio equal to the value of the trigonometric ratio for the angle as found in the table.
4. Solve the resulting equation.

The third side of the triangle can be found by use of the Pythagorean Theorem or by repeating the above process with another of the three trigonometric ratios.

EXERCISE SET 6.4

1. Given a right triangle ABC with the right angle at C.

   (a) Find a if $m(\angle A) = 34°$ and $b = 11$ yd.
   (b) Find a if $m(\angle A) = 52°$ and $b = 5.8$ in.
   (c) Find a if $m(\angle A) = 22°$ and $c = 24$ cm.
   (d) Find a if $m(\angle A) = 46°$ and $c = 10.6$ ft.
   (e) Find b if $m(\angle A) = 45°$ and $a = 40$ in.
   (f) Find b if $m(\angle A) = 26°$ and $a = 31$ yd.
   (g) Find b if $m(\angle A) = 62°$ and $c = 6.5$ m.
   (h) Find b if $m(\angle A) = 11°$ and $c = 4\frac{1}{2}$ in.
   (i) Find c if $m(\angle A) = 58°$ and $a = 10.8$ ft.
   (j) Find c if $m(\angle A) = 81°$ and $a = 3.04$ cm.
   (k) Find c if $m(\angle A) = 77°$ and $b = 15$ mm.
   (l) Find c if $m(\angle A) = 18°$ and $b = 2.03$ m.

2. Given a right triangle ABC with the right angle at C.

   (a) Find $m(\angle A)$ to the nearest degree if $a = 6$ ft and $c = 12$ ft.
   (b) Find $m(\angle A)$ to the nearest degree if $a = 4$ ft and $c = 5\frac{1}{2}$ ft.
   (c) Find $m(\angle A)$ to the nearest degree if $a = 4$ in. and $b = 7$ in.
   (d) Find $m(\angle A)$ to the nearest degree if $a = 3$ cm and $b = 2$ cm.
   (e) Find $m(\angle A)$ to the nearest degree if $b = 12$ cm and $c = 32$ cm.
   (f) Find $m(\angle A)$ to the nearest degree if $b = 4.6$ in. and $c = 7.5$ in.

3. A 20-ft. ladder leaning against the side of a house makes an angle of 25° with the ground. How far is its bottom from the house? How high is its top from the ground?

4. "Caution! Shift Gears—7% grade for the next 5 miles!" is a notice that one reads when descending a mountain road. The seven percent grade means that the ratio of the drop in elevation when divided by the horizontal distance is seven percent. If a person is at an altitude of 10,630 ft when he sees this sign, what is his approximate altitude after he has driven a distance of three miles?

5. A jet takes off from the ground at an angle of 7°. If it maintains this angle for four minutes, how high is it off the ground at the end of four minutes if it travels at the average rate of 180 mph?

6. In a given right triangle ABC with the right angle at C, use the definitions of the trigonometric ratios as applied to this triangle to prove that $\sin A = \cos B$. Pick a specific value for the measure of $\angle A$; from this deduce the measure of $\angle B$ and then verify the result, $\sin A = \cos B$, by checking the specific values in the table of trigonometric ratios.

SUGGESTED READINGS

[8], Ch. 10

[9], Ch. 6

[13], Ch. 12

[20], Ch. 2

[25], Ch. 5

[26], Ch. 10

7

PROOF

We have already discussed the ideas relating to a mathematical system and how it is developed through the process of deductive reasoning. The development begins with undefined words, which are used to make **definitions.** Then the truth of certain statements, called **axioms** or **postulates,** is assumed. By using the rules of logic, together with these definitions and axioms, one next sets about to prove the truth of other statements, called **theorems.** It is the purpose of this chapter to examine each of these basic parts of the system in some detail, with particular emphasis on the rules of logic.

7.1 UNDEFINED WORDS AND DEFINITIONS

Mathematics prides itself on its precise use of language. Wherever possible, words and terms are explicitly defined, so that there is no confusion in their meaning. To prevent confusion, these words are defined in terms of other words whose meanings are clear. But, where are these "other" words defined? Some of them have been defined in terms of still other words.

But eventually, it is *necessary* that the meaning of some words be taken as generally accepted without a definition. A random inspection of a dictionary reveals words that are defined in terms of a second word, which is in turn defined by a third word, which is then defined in terms of the first word. Such a procedure is, of course, being circular but the intent of the lexicographer is not to deceive or confuse, but rather to hope that one of the words being used makes sense to the reader. As an example, the author chose the word "hinder" to start on a journey through the pages of a well-known dictionary. Some of the words used to help establish the meaning of this word were "check," "obstruct," and "prevent." Deciding to pursue "obstruct" as a fruitful example, he found words like "impede" being used. When he looked up "impede," what did he find but "obstruct" and "hinder"!

As an example, look at the following "family" of words:

fascinate	–	bewitch, hold spellbound
spellbound	–	fascinated, entranced
bewitch	–	fascinate, charm, affect by witchcraft
charm	–	a trait that fascinates; magic power
witchcraft	–	practice of magic

Realize that no criticism is intended by this discussion. Quite the contrary! Such a method of operation in defining terms and making ideas clear is necessary. In using undefined words in the definitions of other words, care should be exercised in selecting the undefined words, so that they are as unambiguous as possible.

Mathematicians often choose to define words that have little or no connection to real life situations. It is a matter of pride to feel the sense of accomplishment in creating and inventing with one's own mind a completely abstract system of ideas. Of course, many of the greater inventions in mathematics have served practical needs and have indeed been invented for the purpose of serving these needs. And, what is more interesting, some mathematical systems that were originally devised only as theoretical concepts have found practical application in later generations.

What, then, constitutes a good definition? The following list gives some of the properties of a good definition.

1. It must use words that have already been defined or whose meanings are generally accepted.

EXAMPLE

A triangle is a polygon with three sides. "Polygon" and "sides" are words that have been previously defined or described in this text. The other words are those in common English usage.

2. It must be reversible. If, say, a geometric figure is being defined in terms of certain properties, then the definition tells two things:

   (a) If a geometric object is an instance of the figure being defined, then it has the properties listed in the definition.

   (b) If a geometric object has the properties listed in the definition, then it is an instance of the figure being defined.

EXAMPLE

"A triangle is a polygon with three sides" can also be written as: "A polygon with three sides is a triangle."

3. It should not give more information than is necessary.

EXAMPLE

An isosceles triangle is a triangle with two congruent sides and two congruent angles opposite those sides. This definition gives too much information. Either the two congruent angles or the two congruent sides would be enough to distinguish the isosceles triangle from all other types of triangles. As we have seen, the possession of two congruent sides is sufficient to ensure the possession of two congruent angles (the Isosceles Triangle Theorem) and also that the possession of two congruent angles is sufficient to ensure the possession of the two congruent sides (the converse of the Isosceles Triangle Theorem).

4. It should restrict the term being defined to the smallest class of objects to which it belongs and then give the properties that distinguish it from the other objects of that class.

EXAMPLE

A triangle is a polygon with three sides. Here, "triangle" is being defined as a polygon, rather than as a curve or a closed curve or a simple closed curve. Being a polygon, we know that it is the union of a finite number of segments. The definition distinguishes the triangle from all other polygons by telling us that it has three sides.

5. It should be consistent with the context in which it appears. Thus, it would be incorrect to define a term in such a manner that certain properties would necessarily follow from the definition that would be contradictory to other definitions, postulates, or theorems of the system.

EXAMPLE

A scalesces triangle is an isosceles triangle with its base angles unequal in size.

EXERCISE SET 7.1

1. Which words in the following statements require careful definition or description before the meaning of the statement can be made clear?
   (a) All blondes have fun.
   (b) Tall boys are good athletes.
   (c) Fifty-five was a low score on the test.
   (d) President Kennedy was a good president.
   (e) Colorado is the prettiest state in the United States.
   (f) Driving faster than 60 mph is dangerous.
   (g) Failure to vote in a presidential election is unpatriotic.

2. Criticize the following definitions in terms of the criteria listed above.
   (a) An irrational number is one that is not rational.
   (b) A prime number is one that has no divisors other than itself or 1 and is one of the factors of a composite number.
   (c) A square is a rectangle with two sides congruent.
   (d) A safe driver is a careful driver.
   (e) A scalene triangle is a polygon with three sides that are unequal in length.
   (f) A line is a set of points.
   (g) A novel is a story.
   (h) An astronomer is a person who studies astronomy.

7.2 STATEMENTS

Sentences are certain collections of words. Some sentences are called declarative sentences—a definition is an example of such a sentence. Other sentences ask questions and are called interrogative sentences. There are still others that are called imperative sentences—as such they give a command. "Wow!" or "What a beautiful sight!" are examples of exclamatory sentences. (Some grammarians would not give these latter word collections the status of sentence, for neither contains a verb.)

In the study of logic and proof, we shall be interested in only one kind of sentence, the declarative sentence. Furthermore, such sentences must satisfy two other criteria before they will be considered:

1. The sentence must make sense. Thus, we shall not be interested in such sentences as: "The supricated bojohn chuntled the exalidocious beech sneech"; or "the area of Rhode Island is a rowboat."

2. The sentence must be one that can be assigned the label "true" or "false," but not both. Thus, we shall not consider such sentences as: "He is a nice person."

Sentences that satisfy these criteria will be given the special name of **statement.** Statements and combinations of statements will be the object of study in this chapter. The labels "true" and "false," to be symbolized by T and F, respectively, will be called the "truth value" of a statement. These labels may often seem highly arbitrary and contrary to the way we think things should be.

The arbitrary nature of the labels "true" and "false" comes about for one or both of the following reasons.

1. The study of the rules of logic may be carried out in an abstract, symbolic fashion, whereby the symbols T and F are just elements of the system. As such, they are taken as undefined terms. Even though they may be undefined, their nature becomes clear as the subject develops.

2. Even if the normal, accepted meanings are attached to the labels "true" and "false," the interest of logic is in a conclusion that follows *if* a given statement is asssumed to be true (or false). It may happen that the statement in actual fact may not be what we assume it to be, but that does not matter; we just want to know what *would* happen if it were true (or false).

For the sake of ease of exposition and discussion, as well as printing, statements will be symbolized in this book by lower case Roman letters, such as p, q, or r. For example,

$$\text{"}p: \text{ All cats are lazy."}$$

7.3 NEGATION OF A STATEMENT

Given a statement, it is possible to form a new statement by negating the given one. This negation is effected by inserting the word "not" or the phrase "It is not true that . . ." (or some synonymous words that might make for better sentence structure) closely adjacent to the verb of the statement. This new statement, called the **negation** of the original statement, is symbolized by $\sim p$, where p is the symbol for the original statement.

EXAMPLE

p: Robert is a professional golfer.

$\sim p$:   Robert is not a professional golfer.

The relative truth values of a statement and its negation can best be shown by means of a diagram called a **truth table**.

TRUTH TABLE FOR NEGATION

p	$\sim p$
T	F
F	T

This table tells us that if a statement is considered to be true, then its negation is considered to be false; and if a statement is considered to be false, then its negation is considered to be true.

This idea is quite simple. However, to be sure it is understood and because the analogy to be presented becomes useful in the paragraphs to come, this concept shall be examined from another point of view, namely that of sets.

Let p be "$x \in A$," where A is a set of some objects, say the set of all college students.

Then

$$\sim p \text{ is "} x \notin A \text{."}$$

If we study a Venn diagram of this situation, it can be seen that we are talking about a set and its complement. In Fig. 7-1, point M_1 is shown in set A and point M_2 is shown in the complement of A (symbolized by $\sim A$). Recall the complement of a set A is the set of all those elements in the universal set that are *not* in A.

If A is the set of all college students and if Mike is a person who is represented by M_1, then we say that Mike is a college student; if Mike is

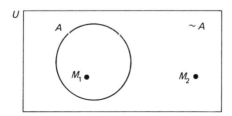

Figure 7-1

represented by M_2, which is not in A, then we say Mike is not a college student.

Before we leave this section, it should be noted that it is possible to use the truth table as a definition for the negation of a statement; that is, the negation of a statement is defined to be a statement that has the properties shown in the table. This mode of thinking has the advantage of being quite precise and leaving no doubt as to the meaning of a negation.

7.4 CONJUNCTION OF STATEMENTS

In the formation of a new statement by negating a previous statement, only one statement was used. In this section and those that follow, new statements will be formed by combining more than one statement.

The first instance is in the **conjunction** of two statements; it is formed by connecting the two with the word "and." If p and q are two statements, their conjunction is symbolized by $p \wedge q$.

EXAMPLE

p: A triangle is a polygon.

q: Topeka is a city in California.

$p \wedge q$:   A triangle is a polygon and Topeka is a city in California.

As the example shows, the new compound statement may very well be a "silly" one in that it is the type of statement one would not ordinarily utter. Yet, it does make sense, even though one of its components seems to be false. We shall be interested in the truth value of this statement, and the truth table given below tells when a conjunction is true. There are two entries, p and q, each with two possible truth values; thus, four lines are required in the table to take care of all possible combinations.

TRUTH TABLE FOR CONJUNCTION

p	q	$p \wedge q$
T	T	T
T	F	F
F	T	F
F	F	F

This table says that a conjunction of two statements is true only when both components are true.

Again, consider the analogy of this situation in set notation. Let A and B be two arbitrary sets.

Let p be "$x \in A$" and

let q be "$x \in B$."

Then $p \wedge q$ is "$x \in A$ and $x \in B$."

The statement, "$x \in A$ and $x \in B$," is the definition of the intersection of two sets A and B. The Venn diagram for this situation is shown in Fig. 7-2.

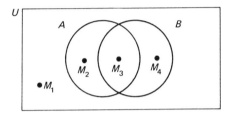

Figure 7-2

If x is at position M_1, then $x \notin A$, $x \notin B$, and $x \notin (A \cap B)$.

Thus, p is false; q is false; $p \wedge q$ is false.

If x is at position M_2, then $x \in A$, $x \notin B$, and $x \notin (A \cap B)$.

Thus, p is true; q is false; $p \wedge q$ is false.

If x is at position M_3, then $x \in A$, $x \in B$, and $x \in (A \cap B)$.

Thus, p is true; q is true; $p \wedge q$ is true.

If x is at position M_4, then $x \notin A$, $x \in B$, and $x \notin (A \cap B)$.

Thus, p is false; q is true; $p \wedge q$ is false.

This analogy agrees exactly with the truth table. It is not a "proof" that the truth table is correct, but it does furnish very convincing evidence that the definition for the conjunction of two statements is a reasonable one that agrees with previous experiences.

The truth table gives the truth values for $p \wedge q$. The truth table for $q \wedge p$ would have the same entries in the last column. Thus, conjunction is a commutative operation on the set of all statements. Rather than write $p \wedge q = q \wedge p$, we have adopted in this text the notation: $p \wedge q \Leftrightarrow q \wedge p$, because $p \wedge q$ and $q \wedge p$ are not "equal" statements (where "equal" means "identical"); they are said to be **equivalent** statements. Two statements, p and q, are equivalent (symbolized by $p \Leftrightarrow q$) if they have the same truth values. This is defined formally by the following truth table.

TRUTH TABLE FOR EQUIVALENCE

p	q	$p \Leftrightarrow q$
T	T	T
T	F	F
F	T	F
F	F	T

Using the truth tables presented thus far, we can construct other truth tables for derived statements that are a combination of the basic kinds of statements.

EXAMPLE

Construct a truth table for $p \wedge \sim q$

p	q	$\sim q$	$p \wedge \sim q$
T	T	F	F
T	F	T	T
F	T	F	F
F	F	T	F

The first two columns are the four possible combinations of truth values for p and q. By using the truth table for negations, we can fill in the third column for $\sim q$. Then if we use the first and third columns and the truth table for conjunction, the entries for the fourth column can be determined.

Truth tables for the conjunction of three or more statements can also be constructed. We shall define $p \wedge q \wedge r$ to be the compound statement $(p \wedge q) \wedge r$.

EXERCISE SET 7.2

1. Write the negation of the following statements.
    (a) The number 2 is a prime number.
    (b) Robert is not a straight-A student.
    (c) California is the most populous state in the Union.
    (d) All athletes are overweight.

   (e) Some teachers are unreasonable.

   (f) No one pays attention to me.

   (g) Somebody up there likes me.

2. Form the conjunction of the pairs of statements given. Try to decide whether the conjunction is true or false.

   (a) The number 147 is a prime number. The number 147 is divisible by 3.

   (b) Geometry is a difficult subject. $\sqrt{676}$ is irrational.

   (c) $\frac{13}{7} > \frac{22}{12}$. $\frac{1}{16} = 0.0625$.

   (d) Philadelphia has more people than Detroit. Polygons are curves.

   (e) Parallel lines never intersect. A square is a rectangle.

3. Complete the following truth table for $(p \wedge q) \wedge r$.

p	q	r	$p \wedge q$	$(p \wedge q) \wedge r$
T	T	T		
T	T	F		
T	F	T		
T	F	F		
F	T	T		
F	T	F		
F	F	T		
F	F	F		

4. Show that conjunction is associative; that is, that $(p \wedge q) \wedge r \Leftrightarrow p \wedge (q \wedge r)$. Do this by completing the following truth table and noting that the last two columns have identical entries.

p	q	r	$p \wedge q$	$q \wedge r$	$(p \wedge q) \wedge r$	$p \wedge (q \wedge r)$
T	T	T				
T	T	F				
T	F	T				
T	F	F				
F	T	T				
F	T	F				
F	F	T				
F	F	F				

5. Construct and complete a truth table for $p \wedge \sim p$.

6. Construct and complete a truth table for $\sim (p \wedge q)$.

7. Construct and complete a truth table for $\sim p \wedge \sim q$.

8. Construct and complete a truth table for $\sim p \wedge q$.

9. Criticize the definition: The statement $p \wedge q \wedge r$ is a statement equivalent to the statement $p \wedge (q \wedge r)$.

7.5 DISJUNCTION OF STATEMENTS

Another compound statement is one that is formed by joining two or more statements by the connective "or." This is called the **disjunction** of statements. Let us examine the case of two statements, since the situation for three or more can be handled in a manner similar to the procedure for conjunctions of statements. In common English usage, the word "or" is used in its exclusive sense—that is, it carries the "either . . . or" connotation. However, in mathematics, the word "or" is used in the inclusive sense, which means "either one or the other or both." The inclusive "or" was used earlier in the definition of the union of two sets. The disjunction of the two statements p and q is symbolized by $p \vee q$.

EXAMPLE

p: Polygons are simple closed curves.

q: The sum of 2 and 2 is 5.

$p \vee q$: Polygons are simple closed curves or the sum of 2 and 2 is 5.

TRUTH TABLE FOR DISJUNCTION

p	q	$p \vee q$
T	T	T
T	F	T
F	T	T
F	F	F

Thus, the disjunction of two statements is false *only* when both statements are false.

The set analog is instructive here. Again, let A and B be two arbitrary sets.

Let p be "$x \in A$" and

let q be "$x \in B$."

Then $p \lor q$ is "$x \in A$ or $x \in B$."

The statement, "$x \in A$ or $x \in B$," is the definition of the union of two sets. The Venn diagram for this is shown in Fig. 7-3.

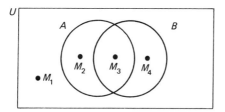

Figure 7-3

As with the conjunction of two statements, let us examine the possibility of x being in each of the four different positions shown in the figure.

If x is at position M_1, then $x \notin A$, $x \notin B$, and $x \notin (A \cup B)$.

Thus, p is false; q is false; $p \lor q$ is false.

If x is at position M_2, then $x \in A$, $x \notin B$, and $x \in (A \cup B)$.

Thus, p is true; q is false; $p \lor q$ is true.

If x is at position M_3, then $x \in A$, $x \in B$, and $x \in (A \cup B)$.

Thus, p is true; q is true; $p \lor q$ is true.

If x is at position M_4, then $x \notin A$, $x \in B$, and $x \in (A \cup B)$.

Thus, p is false; q is true; $p \lor q$ is true.

At the risk of being redundant, we again see how the power of mathematics prevails in giving a perfect analogy. It is hoped that the reader is convinced that the truth table for the disjunction of two statements is a reasonable one.

The truth tables for the disjunction of three or more statements are constructed in a manner identical to that for conjunctions, since $p \lor q \lor r$ is defined to be $(p \lor q) \lor r$.

EXERCISE SET 7.3

1. Form the disjunction of the statements given below. Decide on the truth of these disjunctions.
   (a) The number 147 is a prime number. The number 147 is divisible by 3.
   (b) Geometry is a difficult subject. $\sqrt{676}$ is irrational.
   (c) $\frac{13}{7} > \frac{22}{12} \cdot \frac{1}{16} = 0.0625$.
   (d) Philadelphia has more people than Detroit. Polygons are curves.
   (e) Parallel lines never intersect. A square is a rectangle.

2. Construct and complete a truth table for $(p \lor q) \lor r$.

3. Construct a truth table to decide whether an "associative law" holds for the disjunction of statements.

4. Construct and complete a truth table for $p \lor \sim p$.

5. Construct and complete a truth table for $\sim(p \lor q)$.

6. Construct and complete a truth table for $\sim p \lor \sim q$.

7. Construct and complete a truth table for $\sim(p \lor \sim q)$.

8. We have seen that the two connectives, "and" and "or," between statements give the conjunction and disjunction, respectively, of statements and that "commutative" and "associative" laws hold. This question is designed to have you find out whether the two connectives can be used together with a set of statements and have a "distributive" law hold.

   The distributive property for real numbers was given in Chapter 4. From a previous exercise, you should be aware of the two distributive laws that hold for the union and intersection of sets. They are:

$$A \cap (B \cup C) = (A \cap B) \cup (A \cap C)$$

and

$$A \cup (B \cap C) = (A \cup B) \cap (A \cup C)$$

Since union and intersection of sets have been shown to be analogous to disjunction and conjunction of statements, one may suspect that "distributive" laws do hold. To decide whether or not they do, use truth tables to see if:

   (a) The statement $p \lor (q \land r)$ is equivalent to the statement $(p \lor q) \land (p \lor r)$.

   (b) The statement $p \land (q \lor r)$ is equivalent to the statement $(p \land q) \lor (p \land r)$.

7.6   THE IMPLICATION

The compound statement that is most crucial in the study of proof and proof strategy is the **implication.** The implication is of the form: "If p, then q," and is symbolized by $p \Rightarrow q$. The first component, p, is given the name of the **hypothesis** or **antecedent** of the implication, while the second component, q, is called by the name of **conclusion** or **consequent.**

EXAMPLE

> p: The sun is shining.
> q: The birds are singing.
> $p \Rightarrow q$:   If the sun is shining, then the birds are singing.

Since both p and q are meaningful statements in the above example, the implication formed by using p and q is also a meaningful statement. We must decide what the truth value of this and all other implications shall be, given the various possible truth values of the component statements p and q. As with the other derived statements, this can be shown by means of a truth table for implications.

Truth Table for Implication

p	q	$p \Rightarrow q$
T	T	T
T	F	F
F	T	T
F	F	T

Many mathematics students at all levels have been troubled by the first reading of this truth table. Taken together, the first two rows seem reasonable. If we use the statements in the example just given, the first row would read: "If the sun is shining, then the birds are singing," while the second row would read: "If the sun is shining, then the birds are not singing." Certainly both of these compound statements cannot be true, because a person would not know what to conclude if, indeed, the sun was shining. Common sense seems to indicate that the choice is to make the first implication "true," since both of its component parts are true; thus, the second implication is assigned the value of "false."

It is the entries in the last two rows that generally cause the consternation. One way out would be to say that it is simply the definition—a choice had to be made to assign either T or F to the last column entries of those

and the choice was *T*! Essentially, this is the case. But, let us
ne the set analogy to this situation to see why the choice was a
nable one, and, in fact, the only one that could be made.

Let *p* be "*x* ∈ *A*" and

let *q* be "*x* ∈ *B*."

Then *p* ⇒ *q* is "If *x* ∈ *A*, then *x* ∈ *B*."

"If *x* ∈ *A*, then *x* ∈ *B*" is the definition of the subset relationship,
A ⊆ *B*. The Venn diagram for the subset relationship, *A* ⊆ *B*, is shown
in Fig. 7-4.

In this diagram, there are only three possible positions that *x* could
occupy. Let us examine each in turn.

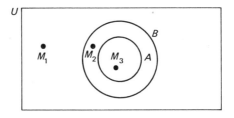

Figure 7-4

If *x* is at position M_1, then *x* ∉ *A*, *x* ∉ *B*, but *A* ⊆ *B*.

Thus, *p* is false; *q* is false; but *p* ⇒ *q* is **true.**

If *x* is at position M_2, then *x* ∉ *A*, *x* ∈ *B*, but *A* ⊆ *B*.

Thus, *p* is false; *q* is true; but *p* ⇒ *q* is **true.**

If *x* is at position M_3 then *x* ∈ *A*, *x* ∈ *B*, and *A* ⊆ *B*.

Thus, *p* is true; *q* is true; and *p* ⇒ *q* is **true.**

We have one more case to examine: *x* ∈ *A* and *x* ∉ *B*. We know
from our knowledge of the subset relationship that if such a situation existed,
A could not be a subset of *B*. This fourth case cannot be shown in Fig. 7-4,
because that figure *does* show *A* as a subset of *B*. If we wished to show
the fourth case in a diagram, we would have to exhibit one in which *A* was

not a subset of *B*. In other words, if $x \in A$ (which is p) were true and $x \in B$ (which is q) were false, then $A \subseteq B$ (which is $p \Rightarrow q$) would be false. This set analogy agrees exactly with the truth table defining the implication $p \Rightarrow q$.

When an implication formed from the statements p and q is given, it is also possible to form other implications by using the negations of these statements and the original statements themselves. Some of the important implications have special names that are used in reference to the original implication $p \Rightarrow q$. Given the implication $p \Rightarrow q$, we can form

(a) the **converse** of the implication: $q \Rightarrow p$

(b) the **inverse** of the implication: $\sim p \Rightarrow \sim q$

(c) the **contrapositive** of the implication: $\sim q \Rightarrow \sim p$

The truth tables for these three are given in the exercises that follow. When doing these exercises, check for the equivalence of these derived implications with respect to each other and to the original implication.

EXERCISE SET 7.4

1. Put the following into the "if-then" form of an implication:
    (a) All athletes are overweight.
    (b) Martians are from outer space.
    (c) All horses have four legs.
    (d) All freshmen are serious students.
    (e) I get heartburn whenever I eat tamales.
    (f) A square is a rectangle.

2. Form the converse, inverse, and contrapositive for each of the following implications:
    (a) If it stops raining, we will go on a picnic.
    (b) If you do all the exercises, you will pass this course.
    (c) If you do not eat your spinach, then you will not grow up to be strong like Popeye.
    (d) If you drive carefully, you will not have an accident.
    (e) If he does not stop eating sweets, he will get fat.
    (f) I shall be unhappy if I fail this test.
    (g) If a number is prime, then it is not composite.
    (h) If snow is green, then Christmas is in July.
    (i) If something is not worth doing, it is not worth doing well.

3. Complete the following truth table for the converse of the implication $p \Rightarrow q$.

p	q	$q \Rightarrow p$
T	T	
T	F	
F	T	
F	F	

4. Complete the following truth table for the inverse of the implication $p \Rightarrow q$.

p	q	$\sim p$	$\sim q$	$\sim p \Rightarrow \sim q$
T	T			
T	F			
F	T			
F	F			

5. Complete the following truth table for the contrapositive of the implication $p \Rightarrow q$.

p	q	$\sim p$	$\sim q$	$\sim q \Rightarrow \sim p$
T	T			
T	F			
F	T			
F	F			

6. Construct and complete a truth table for $p \Rightarrow \sim p$.

7. Construct and complete a truth table for $\sim (p \Rightarrow q)$. This negation is equivalent to another derived statement involving p and q and negations thereof. Check either your intuition, your knowledge, or one of the previous exercises or examples of this chapter to find out which one it is. Check your finding by substituting reasonable and sensible statements for p and q.

8. Construct and complete a truth table for $(p \Rightarrow q) \wedge (q \Rightarrow p)$. When done, check your result with the truth table for equivalence of statements p and q given earlier in this chapter.

9. Construct and complete a truth table for $(p \Rightarrow q) \Rightarrow q$.

7.7 TAUTOLOGIES

We have been looking at a lot of truth tables so far in this chapter. By and large, almost all of them belong in the category of "iffy" statements, in that the statements are sometimes true and sometimes false, depending upon the relative truth values of the simple statements that were the component parts of the more complex statements. To be sure whether a particular statement was true, a person would have to work out the truth tables or commit them to memory. Such a situation is not ideal, but, of course, we have to live with it. Fortunately, there are some statements that are *always* true, regardless of the relative truth values of the component parts. Such statements are called **tautologies**; the more important ones happen to be the very rules of logic that we seek. These are used so often that the reader should commit them to memory.

The first is a simple one. It is Exercise 3 of Exercise Set 7.3. It is $p \lor \sim p$, which goes under the name of the Law of the Excluded Middle. Its truth table is repeated for emphasis.

LAW OF THE EXCLUDED MIDDLE

($p \lor \sim p$ *is always true.*)

p	$\sim p$	$p \lor \sim p$
T	F	T
F	T	T

It is also important to note the statement $p \lor \sim p$ is equivalent to $\sim (p \lor \sim p)$. Put into words, the equivalence of these two statements says that given a statement, either it or its negation must be true but both cannot be true at the same time.

The next tautology tells what we must have in order to use an implication correctly. If you did Exercise 9 from the last set, you saw that an implication $p \Rightarrow q$ is not enough to guarantee the truth of q. (The last entry of F in the final column tells you that.) What is needed is the implication $p \Rightarrow q$ and the statement p in order to be certain of the truth of q. In other words the statement

$$[(p \Rightarrow q) \land p] \Rightarrow q \text{ is a tautology}$$

as the truth table shows. This rule is called the Rule of Detachment or sometimes by the Latin name. *Modus Ponens.*

RULE OF DETACHMENT

p	q	$p \Rightarrow q$	$(p \Rightarrow q) \wedge p$	$[(p \Rightarrow q) \wedge p] \Rightarrow q$
T	T	T	T	T
T	F	F	F	T
F	T	T	F	T
F	F	T	F	T

As you might expect, the Rule of Detachment is most important when we set out to write a proof. Rather than try to give an example at this point, we shall postpone an example until a later section of the chapter.

Another tautology of great importance in deductive thinking is the Law of Syllogism, which states: $[(p \Rightarrow q) \wedge (q \Rightarrow r)] \Rightarrow (p \Rightarrow r)$. Note that this law has a sort of transitive property about it. Here is the truth table.

LAW OF SYLLOGISM

p	q	r	$p \Rightarrow q$	$q \Rightarrow r$	$(p \Rightarrow q) \wedge (q \Rightarrow r)$	$p \Rightarrow r$	$[(p \Rightarrow q) \wedge (q \Rightarrow r)] \Rightarrow (p \Rightarrow r)$
T	T	T	T	T	T	T	T
T	T	F	T	F	F	F	T
T	F	T	F	T	F	T	T
T	F	F	F	T	F	F	T
F	T	T	T	T	T	T	T
F	T	T	T	F	F	T	T
F	F	T	T	T	T	T	T
F	F	F	T	T	T	T	T

Because it will be important in what follows, let us look again at the equivalence of statements, $p \Leftrightarrow q$. Recall that $p \Leftrightarrow q$ is a true statement only when p and q have the same truth values. It can be shown by a truth table that it is always true—a tautology—that $(p \Leftrightarrow q) \Rightarrow [(p \Rightarrow q) \wedge (q \Rightarrow p)]$. You may very well be tired of looking at truth tables by now; therefore, it has been left as an exercise. Also, it is always true that $[(p \Rightarrow q) \wedge (q \Rightarrow p)] \Rightarrow (p \Leftrightarrow q)$.

This last paragraph points out the reason for the symbol "$\Leftrightarrow$" for equivalence. It has been shown that $p \Leftrightarrow q$ means $p \Rightarrow q$ and $q \Rightarrow p$. If we write the letters in the same order for p and q, this second implication would be written $p \Leftarrow q$. Combining $p \Rightarrow q$ and $p \Leftarrow q$ gives the single statement $p \Leftrightarrow q$.

From the exercises and examples of this chapter, you should have observed that there are many statements that are equivalent. It is not necessary to memorize all these equivalences, with one exception. Because of its utility in proving the truth of implications, you should remember:

An implication is equivalent to its contrapositive;

that is: $(p \Rightarrow q) \Leftrightarrow (\sim q \Rightarrow \sim p)$.

This fact could be shown by a truth table, but rather than resorting to that device, let us appeal to common sense. Consider the implication $p \Rightarrow q$; then consider an implication with $\sim q$ as the hypothesis and either p or $\sim p$ as the conclusion—that is, look at $\sim q \Rightarrow p$ and $\sim q \Rightarrow \sim p$. If we consider only those cases where $\sim q$ is true, then these two implications are negations of each other, and, therefore, by the Law of the Excluded Middle, both cannot be true, but one of them must be. If $\sim q \Rightarrow p$ is the one that is true, then we can combine this implication with $p \Rightarrow q$, and by using the Law of Syllogism, come up with $\sim q \Rightarrow q$, which is impossible to accept if $\sim q$ is true. Thus, we are forced to reject the implication $\sim q \Rightarrow p$ and thereby accept $\sim q \Rightarrow \sim p$ as true.

The equivalence of an implication and its contrapositive is often put into short statements that are seen in mathematical literature. The implication $p \Rightarrow q$ leads to verbal statements such as:

If p, then q.
q if p.
p is sufficient for q.
p implies q.
q is implied by p.

Because of the equivalence of $\sim q \Rightarrow \sim p$ to the above, we note that if we do not have q, then we do not have p. This fact is often translated into one of the two following verbal statements:

p only if q.
q is necessary for p.

Note that this equivalence of an implication and its contrapositive must not be confused with an implication and its converse. The converse of an implication $p \Rightarrow q$ is the implication $q \Rightarrow p$, which is not necessarily true if the original implication is true. However, if $p \Rightarrow q$ and $q \Rightarrow p$ are both true, then this is the same thing as saying $p \Leftrightarrow q$. This fact is often expressed by the following:

p is equivalent to q.

p if and only if q (often shortened to p iff q).

p is necessary and sufficient for q.

A necessary and sufficient condition for q is p.

It is worth noting in closing this section that all definitions are in the category of being "if and only if" statements. This was referred to as the property of reversibility in the earlier section of this chapter.

EXERCISE SET 7.5

1. Show by means of a truth table that

$$(p \Leftrightarrow q) \Leftrightarrow [(p \Rightarrow q) \wedge (q \Rightarrow p)]$$

is a tautology.

2. Verify by truth tables that the following are tautologies:
   (a) $\sim (\sim p) \Rightarrow p$
   (b) $(p \wedge q) \Rightarrow p$
   (c) $[\sim q \wedge (p \Rightarrow q)] \Rightarrow \sim p$
   (d) $[\sim q \wedge (p \vee q)] \Rightarrow p$
   (e) $[(p \wedge q) \Rightarrow r] \Rightarrow [p \Rightarrow (q \Rightarrow r)]$
   (f) $(p \Rightarrow q) \Leftrightarrow [\sim p \vee q]$

3. "Translate" each of the following verbal statements into the "If . . . , then . . ." form of the statement.
   (a) Studying is a necessary condition for passing this course.
   (b) In order for one to remain thin, it is sufficient to avoid eating potatoes.
   (c) Watering plants is sufficient to make them grow.
   (d) I shall lend you money only if you promise to pay it back.
   (e) Congruence of triangles implies similarity of triangles.
   (f) A sufficient condition that a polygon be a rectangle is that it is a square.
   (g) Only if I love you will I marry you.
   (h) A necessary condition that a triangle be isosceles is that it have two congruent angles.
   (i) The fact that two triangles have two pairs of corresponding angles respectively congruent is sufficient to guarantee that their third angles will also be congruent to each other.
   (j) To have fun, it is necessary to be rich.

4. Write out the two implications in the "If . . . , then . . ." form that are contained in the following statements.
   (a) $a + c = b + c$ if and only if $a = b$.

(b) A triangle is equilateral if and only if it is equiangular.

(c) A necessary and sufficient condition that the integer b is a divisor of the integer a is that a be a multiple of b.

(d) Definition: A scalene triangle is one with no two sides congruent.

(e) Being a human is equivalent to being a descendant of Adam and Eve.

5. Symbolize the following statements by using the negation symbol and the implication symbol as appropriate.

(a) If Rock Hunter uses "All-White" toothpaste, then he will not have cavities.

    p: Rock Hunter uses "All-White" toothpaste.

    q: He will have cavities.

(b) Having cavities is a sufficient condition for going to the dentist.

    q: He has cavities.

    r: He will go to the dentist.

(c) A necessary condition for saving money is not going to the dentist.

    r: He will go to the dentist.

    t: He will save money.

(d) Rock Hunter will save money by not using "All-White" toothpaste.

    p: Rock Hunter uses "All-White" toothpaste.

    t: He will save money.

(e) If the moon is made of blue cheese, then astronauts will land on it.

    p: The moon is made of blue cheese.

    q: Astronauts will land on the moon.

(f) A necessary condition that the astronauts will land on the moon is that they will be able to rocket off the moon's surface.

    q: Astronauts will land on the moon.

    r: Astronauts will be able to rocket off the moon's surface.

(g) Astronauts will be able to rocket off the moon's surface only if they are not overweight.

    r: Astronauts will be able to rocket off the moon's surface.

    t: They are overweight.

(h) Eating blue cheese implies being overweight.

    t: They are overweight.

    s: Eating blue cheese.

(i) If lines are parallel, then they do not intersect.

    p: Lines are parallel.

    q: Lines intersect.

(j) Being skew is sufficient for lines to not intersect.

    q: Lines intersect.

    s: Lines are skew.

7.8   PROOF

Now that we have at our disposal the rules of logic, we should examine how to put them to use. In the development of a mathematical subject, the desire is to demonstrate unequivocally the truth of certain statements or propositions. These propositions are the theorems, the conjecturing and hypothesizing of which come from experimentation, induction, and a combination of other factors that may be called good mathematical sense. Basically, the proof of the theorems uses the Rule of Detachment and the Law of Syllogism and combinations thereof, together with the free substitution of equivalent statements whenever necessary. Both of these rules require the knowledge of the truth of one or more implications. These implications are the previously proven theorems, or in the case of the very beginnings of the development, the postulates and the definitions.

Let us examine a hypothetical case first, so that we can look at the structure of a direct proof; then, in succeeding paragraphs, we can examine specific instances. Suppose that we wish to prove the theorem, $p \Rightarrow q$, to be a true statement. Using only the Law of Syllogism, we might proceed as follows:

(a) Use a set of known-to-be-true implications, the last of which has q as its conclusion. Symbolically, these might be $p \Rightarrow r_1$, $r_1 \Rightarrow r_2$, $r_2 \Rightarrow r_3$, $r_3 \Rightarrow q$. (Of course, there may be fewer or more than what are listed here, but this is only an example.)

(b) Use the Law of Syllogism repeatedly as shown in the schematic diagram.

$$
\left.
\begin{array}{c}
p \Rightarrow r_1 \\
\wedge \\
r_1 \Rightarrow r_2
\end{array}
\right\} \Rightarrow (p \Rightarrow r_2)
$$

$$
\left.
\begin{array}{c}
\wedge \\
r_2 \Rightarrow r_3
\end{array}
\right\} \Rightarrow (p \Rightarrow r_3)
$$

$$
\left.
\begin{array}{c}
\wedge \\
r_3 \Rightarrow q
\end{array}
\right\} \Rightarrow (p \Rightarrow q)
$$

Any of the known-to-be-true implications may be the contrapositive of another implication; one can even prove the contrapositive of the given to-be-proven implication if doing so is more convenient. The choice of the intermediate implications (theorems, postulates, definitions) is made on the basis of the theorem to be proven, what other statements are available, and the skill and ingenuity of the person writing the proof The final implication (in this example, it is $r_3 \Rightarrow q$) must have as its conclusion the simple statement that is the conclusion of the theorem. In other words, if you are trying to prove the theorem $p \Rightarrow q$, the final implication used in the proof does not have the form $r \Rightarrow t$! Once you have chosen the final intermediate implication, say $r_3 \Rightarrow q$, out of the several that are possible (such as $r_1 \Rightarrow q$, $r_2 \Rightarrow q$, $r_3 \Rightarrow q$, $r_4 \Rightarrow q$), then the preceding implication must be one that has r_3 as its conclusion, and so on. Thus, any analysis or planning of a proof must start at the end, rather than at the first step.

If the Rule of Detachment is used, we must again choose some intermediate implications, using the same guidelines as outlined in the previous paragraph. But here, the strategy, or underlying rationale, is a bit different. In most instances, it is the desire to show the truth of an implication $p \Rightarrow q$ with the added proviso that p is true; we do not care what happens when p is false. Thus, thinking of the defining truth table for the implication $p \Rightarrow q$, we can reason that all that is needed to prove $p \Rightarrow q$ true, with the proviso of p being true, is to show that q is true.

Schematically, the proof would look like this:

$$
\left.
\begin{array}{c}
p \Rightarrow r_1 \\
\wedge \\
p
\end{array}
\right\} \Rightarrow r_1
\left.
\begin{array}{c}
\\
\wedge \\
r_1 \Rightarrow r_2
\end{array}
\right\} \Rightarrow r_2
\left.
\begin{array}{c}
\\
\wedge \\
r_2 \Rightarrow r_3
\end{array}
\right\} \Rightarrow r_3
\left.
\begin{array}{c}
\\
\wedge \\
r_3 \Rightarrow q
\end{array}
\right\} \Rightarrow q
$$

These rules of logic are used in the proofs of geometry theorems. Before an example of a familiar theorem is given, several abstract "theorems" will be given to further demonstrate the structure of a proof.

EXAMPLES

(a) Hypothesis: $p \Rightarrow q$, $q \Rightarrow \sim t$, $\sim r \Rightarrow t$, p.

Prove: r

(If this were the statement of a theorem to be proved, then the problem would read: "Prove the truth of the theorem $p \Rightarrow r$." Under these conditions, one assumes the truth of p. The other implications stated in the hypothesis of this example would be previously accepted-as-true implications, such as postulates, definitions, or previously proven theorems.)

Proof:

Statement	Reason
1. $p, p \Rightarrow q$	1. Hypothesized as true
2. $\therefore q$	2. Statement 1 and the Rule of Detachment
3. $q \Rightarrow \sim t$	3. Hypothesized as true
4. $\therefore \sim t$	4. Statements 2 and 3 and the Rule of Detachment
5. $\sim r \Rightarrow t$	5. Hypothesized as true
6. $\sim t \Rightarrow r$	6. Statement 5 and the Rule of Contrapositive Equivalence
7. $\therefore r$	7. Statements 4 and 6 and the Rule of Detachment

(*Note:* the symbol $\therefore$ stands for the word "therefore.")

(b) Hypothesis:

(1) A necessary condition for Frenchmen to not eat broiled snails is that dogs do not have four legs.

(2) If baseballs are round, then dogs have four legs.

(3) If Frenchmen eat broiled snails, then Popeye does not eat spinach.

(4) Baseballs are round.

Prove: Popeye does not eat spinach.

Proof: Symbolize each simple statement by a letter. Thus,

p: Frenchmen eat broiled snails.
q: Dogs have four legs.
r: Baseballs are round.
s: Popeye eats spinach.

Using these symbols, the hypothesized-as-true statements become:

(1) $\sim p \Rightarrow \sim q$; (2) $r \Rightarrow q$; (3) $p \Rightarrow \sim s$; (4) r, and the conclusion to be proved is $\sim s$.

In the reasons for each statement, only the number of the previously given statement will be written, along with the reason or rule of logic.

Statement	Reason
1. r, $r \Rightarrow q$	1. Hypothesized as true
2. $\therefore$ q	2. 1, Rule of Detachment
3. $\sim p \Rightarrow \sim q$	3. Hypothesized as true
4. $\therefore$ $q \Rightarrow p$	4. 3, Rule of Contrapostive Equivalence
5. $\therefore$ p	5. 2, 4, Rule of Detachment
6. $p \Rightarrow \sim s$	6. Hypothesized as true
7. $\therefore$ $\sim s$	7. 5, 6, Rule of Detachment

This particular mode of analysis will now be applied to the familiar Isosceles Triangle Theorem. Since the proof has already been given in Chapter 5, no further discussion of the strategy in "figuring out" the proof is necessary. The theorem will be stated in slightly different terms than in Chapter 5, so that the present exposition will be facilitated.

THEOREM (Isosceles Triangle Theorem)

In $\triangle ABC$, if $\overline{AB} \cong \overline{AC}$, then $\angle C \cong \angle B$.

Proof: Refer to Fig. 7-5 and consider the correspondence $ABC \leftrightarrow ACB$.

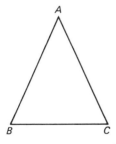 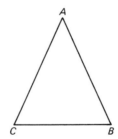

Figure 7-5

Each of the statements needed in the proof shall be symbolized by the following symbols:

p: $\overline{AB} \cong \overline{AC}$
q: $\overline{AC} \cong \overline{AB}$
r: $\angle A \cong \angle A$

$$s: \quad \triangle ABC \cong \triangle ACB$$
$$t: \quad \angle C \cong \angle B$$

The implications needed to build the logical chain are not hypothesized-as-true, as in the previous examples, but are previously established results. The ones used in the proof, together with the symbolic notation derived from the above symbols, are:

(1) The symmetric property of segment congruence. $(p \Rightarrow q)$
(2) The SAS Postulate. $([p \wedge r \wedge q] \Rightarrow s)$
(3) The definition of triangle congruence. $(s \Rightarrow t)$

Following is the proof in a form much expanded from that presented in Chapter 5.

Statement	Symbol	Reason
1. $\overline{AB} \cong \overline{AC}$	p	1. Hypothesized as true
2. $\angle A \cong \angle A$	r	2. Reflexivity of angle congruence
3. Symmetry of congruence	$p \Rightarrow q$	3. Previously defined
4. $\therefore \overline{AC} \cong \overline{AB}$	$\therefore q$	4. 1, 3, Rule of Detachment
5. 1, and 2, and 4	$p \wedge r \wedge q$	5. 1, 2, 4, and truth table definition of conjunction
6. SAS Postulate	$[p \wedge r \wedge q] \Rightarrow s$	6. Previously postulated
7. $\therefore \triangle ABC \cong \triangle ACB$	$\therefore s$	7. 5, 6, Rule of Detachment
8. Definition of triangle congruence	$s \Rightarrow t$	8. Previously defined
9. $\therefore \angle C \cong \angle B$	$\therefore t$	9. 7, 8, Rule of Detachment

Such an expanded form is not typically given in a proof; as such it was given for illustrative purposes. In common practice, the proper use of the Rule of Detachment or any of the other rules of logic is assumed. The reason given in the "Reason" column is usually one of the previously established implications. The rules required for the contracted form of the proof are:

(1) If the "Reason" is an implication, then the hypothesis of that implication must have been established as true in one of the previous "Statements."

(2) If the "Reason" is an implication, then the conclusion of that implication must be a statement that is essentially identical to the "Statement" corresponding to that "Reason."

As an example of these rules, the proof of the Isosceles Triangle Theorem that was presented in Chapter 5 will be reproduced, with the appropriate symbolism for the statements agreed upon in the above illustration written in parentheses.

Statement	Reason
1. $\overline{AB} \cong \overline{AC}$ (p)	1. Hypothesized as true
2. $\angle A \cong \angle A$ (r)	2. Reflexive property of angle congruence
3. $\therefore \overline{AC} \cong \overline{AB}$ (q)	3. Symmetric property of segment congruence $(p \Rightarrow q)$
4. $\therefore \triangle ABC \cong \triangle ACB$ (s)	4. SAS Postulate $([p \wedge r \wedge q] \Rightarrow s)$
5. $\therefore \angle C \cong \angle B$ (t)	5. Definition of triangle congruence $(s \Rightarrow t)$

7.9   INDIRECT PROOF

The renowned fictional detective, Sherlock Holmes, would often establish the guilt of a person in a set of suspected persons, not by directly showing that the person was guilty, but rather by showing that all the other suspects were innocent. He used a process of elimination, eliminating all possibilities until only one remained. This method of proof has application in the field of mathematics and is known as an **indirect proof**.

Suppose we wish to prove the implication $p \Rightarrow q$ true, again with the added proviso that p is assumed to be true. Ordinarily, we would proceed by the methods of the previous section to prove q true. However, if it became difficult to build up a chain of implications to show the truth of q, then the strategy would be to try an indirect proof. The strategy rests upon the Law of the Excluded Middle.

By the Law of the Excluded Middle, there are only two possibilities with respect to the statement q: Either q is true or $\sim q$ is true; only one of them may be true, but one of them *must* be true. The plan of the indirect proof, then, is to show $\sim q$ to be false; that is, to show $\sim (\sim q)$ true. Doing so shows that q is true since $q \Leftrightarrow \sim (\sim q)$.

We begin the indirect proof by assuming the truth of $\sim q$ (this is called the **indirect proof assumption**) and then, by the methods of the previous section, showing that this assumption leads to the contradiction of some known fact. This known fact may be the hypothesis of the theorem to be proven, although it need not be. If we let this known fact be symbolized by t, then this entire discussion may be symbolized by the following statement, which is known as the **Indirect Proof Rule:**

$$[(\sim q \Rightarrow \sim t) \wedge t] \Rightarrow \sim (\sim q)$$

This rule is a tautology. Rather than show that it is always true by means of a truth table, we can demonstrate it by noting that $(\sim q \Rightarrow \sim t)$ $\Leftrightarrow (t \Rightarrow q)$ by the contrapositive equivalence rule and that $\sim (\sim q) \Leftrightarrow q$, as mentioned above. Substituting these equivalent statements into the Indirect Proof Rule, we get the familiar Rule of Detachment, which is known to be a tautology.

EXAMPLES

(a) Hypothesis: $p \Rightarrow r, \sim t \Rightarrow \sim r, t \Rightarrow q, p$

Prove: q

After stating the hypothesized-as-true statements, the indirect proof assumes the negation of what is to be proved as true.

Statement	Reason
1. $p \Rightarrow r$.	1. Hypothesized as true
2. $\sim t \Rightarrow \sim r$	2. Hypothesized as true
3. $t \Rightarrow q$	3. Hypothesized as true
4. p	4. Hypothesized as true
5. $\sim q$	5. Indirect proof assumption
6. $\sim q \Rightarrow \sim t$	6. 3, Rule of Contrapositive Equivalence
7. $\therefore \sim t$	7. 5, 6, Rule of Detachment
8. $\therefore \sim r$	8. 2, 7, Rule of Detachment
9. $\sim r \Rightarrow \sim p$	9. 1, Rule of Contrapositive Equivalence
10. $\therefore \sim p$	10. 8, 9, Rule of Detachment
11. $p \wedge \sim p$	11. 4, 10, Definition of Conjunction
12. But, $p \wedge \sim p$ is a contradiction.	12. Law of Excluded Middle
13. $\therefore \sim (\sim q)$	13. Indirect Proof Rule
14. $\therefore q$	14. $q \Leftrightarrow \sim (\sim q)$

(b) This is the same example as example (a) of the previous section. The proof given will be an indirect one. The hypothesis, restated, is: $p \Rightarrow q$, $q \Rightarrow {\sim}t$, ${\sim}r \Rightarrow t$, p, and the proof is to show r true.

Statement	Reason
1. $p, p \Rightarrow q$	1. Hypothesized as true
2. $q \Rightarrow {\sim}t$	2. Hypothesized as true
3. ${\sim}r \Rightarrow t$	3. Hypothesized as true
4. ${\sim}r$	4. Indirect proof assumption
5. $\therefore t$	5. 3, 4, Rule of Detachment
6. q	6. 1, Rule of Detachment
7. $\therefore {\sim}t$	7. 2, 6, Rule of Detachment
8. $\therefore {\sim}t \wedge t$	8. 5, 7, Definition of Conjunction
9. But, ${\sim}t \wedge t$ is a contradiction.	9. Law of the Excluded Middle
10. $\therefore {\sim}({\sim}r)$	10. Indirect Proof Rule
11. $\therefore r$	11. $r \Leftrightarrow {\sim}({\sim}r)$

None of the proofs previously presented in Chapters 5 and 6 will be given as examples of an indirect proof, since all were given in a direct manner. To try to convert one of them into an indirect proof would result in a contrived, and perhaps confusing, example. Many of the proofs in the following chapter, however, will be indirect proofs.

EXERCISE SET 7.6

1. In the following, prove the truth of the statement given in the "Prove" portion.

   The statements given as hypotheses may be assumed to be true.

   Use both a direct proof and an indirect proof.

   (a) Hypothesis: $p, p \Rightarrow r, {\sim}t \Rightarrow {\sim}r, t \Rightarrow q$

   Prove: q

   (b) Hypothesis: $s, {\sim}p \Rightarrow q, q \Rightarrow {\sim}s$

   Prove: q

   (c) Hypothesis: $p \Rightarrow {\sim}q, t \Rightarrow q, {\sim}t \Rightarrow {\sim}r, r$

   Prove: ${\sim}p$

(d) Hypothesis: $\sim p \Rightarrow \sim t, \sim m \Rightarrow t, m \Rightarrow \sim s, s$

Prove: p

2. In the following, decide upon the validity of the stated conclusion—
   that is, decide if it is a true statement arrived at by correct usage of the
   laws of logic based upon the stated hypotheses.

   (a) Hypothesis: All athletes are overweight.
                   Walter is an athlete.

   Conclusion: Walter is overweight.

   (b) Hypothesis: All members of the Swanksville Country Club are
                   wealthy persons.
                   I. Gotta Lotsadough is a wealthy person.

   Conclusion: I. Gotta Lotsadough belongs to the Swanksville
               Country Club.

   (c) Hypothesis: All cars are bars.
                   All bars are wooden.

   Conclusion: All cars are wooden.

   (d) Hypothesis: If snakes will not slither, then monkeys will not chatter.
                   A sufficient condition that monkeys will chatter is that
                   tigers are ferocious.
                   A necessary condition that snakes will slither is that
                   weeds will wiggle.
                   Tigers are ferocious.

   Conclusion: Weeds wiggle.

   (e) Hypothesis: If movie stars are glamorous, they do not need makeup.
                   Movie stars do not need makeup only if they get up
                   early.
                   If movie stars' teeth do not sparkle, then they do not
                   get up early.
                   Movie stars are glamorous.

   Conclusion: Movie stars' teeth do not sparkle.

3. Here is a classical problem that has appeared in many books. Its
   solution requires the use of some indirect reasoning. Can you solve it?

   Glass, Stone, and Anderson are the engineer, brakeman, and fireman on
   a train, although not necessarily in that order. Riding on the train are
   three passengers with the same last names, to be identified in what
   follows by a "Mr." before their names.

(a) Mr. Anderson lives in San Francisco.

(b) The brakeman lives in Denver.

(c) Mr. Stone long ago forgot all the algebra he learned in high school.

(d) The passenger whose name is the same as the brakeman's lives in St. Louis.

(e) The brakeman and one of the passengers, a distinguished mathematician, attend the same church.

(f) Glass beat the fireman at table tennis.

Who is the engineer?

SUGGESTED READINGS

[8] Ch. 1

[15] Ch. 4

[22] Ch. 2

[26] Ch. 15

[29] Ch. 1

8

PARALLELS
AND PARALLELOGRAMS

In this chapter, we shall return to subject matter that belongs strictly to the field of geometry, namely, the study of parallel lines. Now that we have some of the tools of formal logic at our disposal, we shall be more rigorous in our development of the subject, without, however, abandoning intuition entirely. In fact, let us begin this chapter with a section that may contradict some of your intutitive feelings about lines. The first section appears to be out of place in a chapter on parallel lines; it is included to motivate you to think about a very fundamental question that is asked at the end of the third paragraph of that section.

8.1 GEOMETRY ON A SPHERE

Up to now, we have been studying points, lines, and planes, which seem to exist in a space that goes on forever. But, suppose we restrict ourselves to points on a sphere. Our own earth is almost spherical in shape, so that this example is by no means a far-fetched one. A sphere is a set of points, all of which are the same distance from some fixed point, called its center. This distance is called the radius of the sphere.

Consider two points, A and B, on this sphere. We want to think what it means to have a line connecting these two points, much the same as we have had thus far in our ordinary geometry. How shall we go about it?

Well, imagine the ordinary straight line $\overleftrightarrow{AB}$ connecting these two points.

178

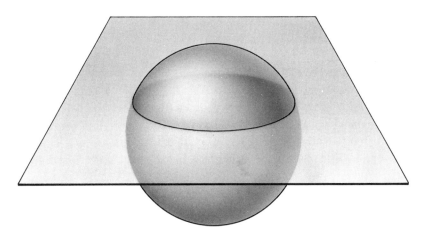

Figure 8-1

It is a line that "pierces through" the sphere and has only the points A and B in common with the sphere. Now, consider any one of the infinite planes lying on line $\overleftrightarrow{AB}$. The intersection of any one of these planes with the sphere will be a set of points that is a circle. One such plane is shown in Fig. 8-1. It would seem sensible, then, to describe the *line* containing A and B, in the geometry of the set of points that is the sphere, to be the circle containing A and B. However, there is still the question of which circle to use. Each different intersecting plane determines a different circle. Without stretching the imagination too much, it is most logical to choose the plane that contains not only the points A and B, but also the center of the sphere. Then the plane becomes unique, and so does the circle. Such a circle on a sphere is called a **great circle** of the sphere. The equator of the earth is an example of a great circle, providing we assume the earth is a perfect sphere.

If we consider all lines on a sphere as being great circles, then we have to accept as a fact that all lines intersect, and, in fact, intersect not in one point, but in *two* points. Therefore, in the geometry of the sphere, there are no such things as parallel lines as we would like to think of them. Recognize that it would still be possible to define the set of parallel lines as those lines that did not intersect, in the same way as we have done in our ordinary geometry. But, just because we define something does not mean that that something exists. This fact may surprise you, but just because parallel lines are defined in the ordinary geometry that we have been studying does not guarantee the existence of parallel lines. Intuition says there are parallel lines; however, it is necessary to make a statement that parallel lines do exist. This statement can be made by a theorem, or we can postulate the existence of them. Just to test your intuition—what do you suspect? Is it possible to prove the existence of parallel lines or shall we have to postulate their existence?

8.2 SOME PRELIMINARY THEOREMS

Before we look at parallel lines and the facts related to them, it is necessary in the logical development to have several theorems at our disposal. These could have been presented in earlier chapters but have been delayed, so that this development could proceed in a compact fashion. The first result has to do with perpendiculars; it is based upon the discussion in Chapter 4 on the imposition of a coordinate system on a half-plane determined by some line $\overleftrightarrow{AB}$, so that angles could be measured. Recall that this discussion centered upon choosing some point X on a line $\overleftrightarrow{AB}$ to be the vertex of all the possible angles formed. Each real number between 0 and 180 was assigned to a particular unique ray with endpoint X. The interest here centers on the real number 90. This present discussion can be formalized into the intuitively obvious theorem that follows.

THEOREM 8.1

On a fixed point X of a given line l in a plane α, there lies exactly one line m that is perpendicular to l.

Only one comment is necessary. The theorem has to stipulate a given plane, for if that were not included, there would be many such perpendicular lines. Think of the spokes of a wheel, which are all perpendicular to the axis of that wheel at the same point.

Next, we shall prove what is called the Exterior Angle Theorem. If one side of $\triangle ABC$, say $\overline{AB}$, is extended to form ray $\overrightarrow{AB}$, then $\angle DBC$ is determined, where D is some point on $\overrightarrow{AB}$ such that B is between A and D. The angle, $\angle DBC$, is said to be an **exterior angle** of $\triangle ABC$. This is shown in Fig. 8-2. With respect to this exterior angle pictured in Fig. 8-2, $\angle A$ and $\angle C$ are called **remote interior** angles. Since each of the three sides could be extended in either of two directions, each triangle has six exterior angles. The two at each vertex turn out to be congruent to each other, since they are vertical angles.

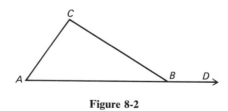

Figure 8-2

180

Let us now state this important, and useful, theorem.

THEOREM 8.2 (Exterior Angle Theorem)

The measure of an exterior angle of a triangle is greater than the measure of either remote interior angle.

Proof: Rather than write out a two-column proof as in previous chapters, the proof is outlined in paragraph form. One of the significant features of this proof is the necessity of introducing extra lines that are not actually given in the statement of the hypothesis of the theorem. These lines are called *construction* lines. Some texts get very technical about the permissibility of drawing such lines and thus make postulates and theorems that allow such lines. We shall not do that here, although we shall be careful not to impose upon such lines conditions that cannot be so imposed. The constructed lines will be illustrated by dashed lines.

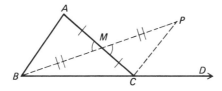

Figure 8-3

Refer to Fig. 8-3. Consider $\triangle ABC$ with exterior $\angle DCA$. The plan is to do half of the proof—to show $m(\angle DCA) > m(\angle A)$.

Point M is found as the midpoint of $\overline{AC}$, so that $\overline{AM} \cong \overline{MC}$, as indicated by the marks of the figure. Segment $\overline{BM}$ is constructed and then extended to a point P such that $\overline{BM} \cong \overline{MP}$, also shown in the figure. $\angle BMA \cong \angle PMC$. (Why?) Thus, $\triangle AMB \cong \triangle CMP$ by the SAS Postulate.

Therefore $\angle BAM \cong \angle PCM$, since they are corresponding parts. The figure seems to indicate that $m(\angle DCM) > m(\angle PCM)$, and, indeed such is the case, although looking at a picture is no proof. The thing that does prove it is the Angle-Addition Postulate, which was Exercise 5 of Exercise Set 6.3. It says

$$m(\angle DCM) = m(\angle DCP) + m(\angle PCM)$$

Simple arithmetic and the rules of inequalities give the inequality we have stated. Since $m(\angle DCM) > m(\angle PCM)$ and since $m(\angle PCM) = m(\angle BAM)$, we conclude $m(\angle DCM) > m(\angle BAM)$, which was what we wanted to prove.

The proof that $m(\angle DCM) > m(\angle ABC)$ is left to the exercises.

One final theorem remains for this section. It is similar to Theorem 8.1, but it deals with a perpendicular from a point X not on a line $\overleftrightarrow{AB}$.

THEOREM 8.3

On a fixed point X not on a given line l, there lies exactly one line m that is perpendicular to l.

This innocent sounding theorem is an interesting one, because it demands a two-pronged proof. First, we must show that there is at least one line— a so-called **existence proof**; then we must show that there is *only* one line —a so-called **uniqueness proof**.

Existence Proof: Refer to Fig. 8-4. This proof relies heavily on construction lines. On line l, choose two distinct points A and B. Draw ray $\overrightarrow{AP}$, forming $\angle PAB$. Now, $\angle PAB$ has a certain measure between 0 and 180. In the other half-plane (the non-P-side) locate a point Q such that $m(\angle QAB) = m(\angle PAB)$. This makes $\angle QAB \cong \angle PAB$, as indicated in the figure. Now, on ray $\overrightarrow{AQ}$, find a point M such that $\overline{AM} \cong \overline{AP}$. Finally, connect P and M to form segment $\overline{PM}$, which we know will interest l, since P and M are in different half-planes. Call this point of intersection N. $\overline{AN} \cong \overline{AN}$, by the Reflexive Property. Thus, $\triangle ANP \cong \triangle ANM$ by the SAS Postulate. As a result, $\angle ANP \cong \angle ANM$. Since these two angles are a linear pair and have equal measures as a result of their congruence, each of them is a right angle. Therefore, $\overline{PM} \perp l$; thus, $\overleftrightarrow{PM}$ is a line through P that is perpendicular to l.

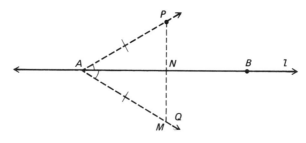

Figure 8-4

Uniqueness Proof: As with most uniqueness proofs, the indirect method of proof is used. We start the proof by assuming there is more than one line through P perpendicular to l and show that this assumption leads to a contradiction.

Suppose there are two lines, m_1 and m_2, both through P and perpendicular to l. If m_1 and m_2 intersect l at points T and R, respectively, we then have $\triangle PTR$. But, the Exterior Angle Theorem says $m(\angle QTP) > m(\angle TRP)$. But both are right angles by construction, so there is a contradiction if we assume the existence of distinct lines m_1 and m_2 perpendicular to l through P.

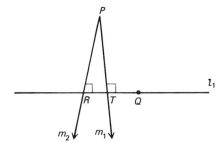

Figure 8-5

Thus, by the Indirect Proof Rule, the negation of the assumption is true; that is, there is exactly one line perpendicular to l through P.

It is important to note that this uniqueness proof and the Exterior Angle Theorem upon which the proof was based make no mention of the sum of the measures of the angles of a triangle.

It is important also to note a special corollary embedded in the uniqueness proof.

COROLLARY 8.4

A triangle cannot have two right angles.

EXERCISE SET 8.1

1. Referring to Fig. 8-3, prove the second part of the Exterior Angle Theorem; that is, show that $m(\angle DCA) > m(\angle ABC)$. (*Hint:* Draw the segment from A to the midpoint of $\overline{BC}$, and then proceed in a similar fashion to the proof in the text. Also think of vertical angles.)

2. How would one measure angles between "lines" on a sphere?

3. Why was the phrase "in a plane" not included in Theorem 8.3, as it was in Theorem 8.1?

4. Suppose you were taking an evening jet from New York to London, and you had a choice of a window seat on either side of the plane. If your choice was based on avoiding the morning sun that would appear as you approach London at the end of the flight, which side would you choose? Explain your reason.

8.3   PARALLEL LINES

We are ready to answer the question posed earlier in this chapter—does the existence of parallel lines have to be postulated or can it be proven as a theorem? We begin by restating the definition of parallel lines.

DEFINITION 8.1

Two lines, m_1 and m_2, are said to be **parallel** if and only if they lie in the same plane, and $m_1 \cap m_2 = \varnothing$.

The symbol for "is parallel to" is "$\|$."   Thus, $m_1 \| m_2$ is read: "Line m_1 is parallel to line m_2."   We shall also say that segments and rays are parallel to each other or to lines if they are subsets of lines that are parallel.

Can you see that the condition, $m_1 \| m_2 = \varnothing$, implies that the two lines must be distinct? This fact keeps the relation of parallelism defined on the set of all lines from being an equivalence relation, for it fails to be reflexive.

The next theorem is used to show that the existence of parallel lines is a fact that can be proven. This theorem sounds very much as if it were related to Corollary 8.4, which it is; but recognize that it says something quite different.

THEOREM 8.5

If two coplanar lines are perpendicular to the same line at two distinct points, then those lines are parallel.

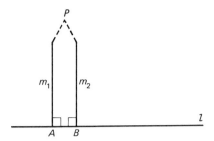

Figure 8-6

Proof: The proof is indirect. Referring to Fig. 8-6, suppose the two lines, m_1 and m_2, are not parallel and meet at some point P. Then point P, together with the points of intersection of m_1 and m_2 with l would form the triangle, $\triangle PAB$. This triangle would have two right angles, which fact contradicts Corollary 8.4. Therefore, the assumption of nonparallelism is false, and we are forced to accept the conclusion that $m_1 \parallel m_2$.

So now we can state the following theorem establishing the existence of parallel lines.

THEOREM 8.6

Through a point P not on a line l, there exists a line m such that $m \parallel l$.

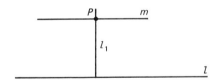

Figure 8-7

Proof: See Fig. 8-7. We are given line l and point P. By Theorem 8.3, there is a line l_1 through P such that $l_1 \perp l$. By Theorem 8.1, there is a line m such that $m \perp l_1$ at point P. Since $m \perp l_1$ and $l \perp l_1$, the preceding theorem, Theorem 8.5, guarantees that $m \parallel l_1$, which is what was to be shown.

The important part to realize is that Theorem 8.6 is an existence theorem that states that there is at least one line through P parallel to the line. What the theorem does not state is the uniqueness of this line. Intuition tells us that this should be the case—that there is exactly one line. For many centuries, mathematicians sought to find a proof of the uniqueness theorem. It was not until the nineteenth century that it was proven that this statement on uniqueness was, in fact, impossible to prove, at least on the basis of the other assumptions made by Euclid. Thus, this statement remains as Euclid's famous Parallel Postulate.

POSTULATE 8.7 (Parallel Postulate)

On a point not on a given line there lies at most one line that is parallel to the given line.

As noted, many centuries passed by without any success in the attempts to "prove" this postulate. Then, when the breakthrough did occur, the amazing thing was that it was discovered by three men who found the result almost at the same time but quite independently of each other. None of the three even knew that the others were working on the same ideas. These three men were Nicolai Lobachevski, a Russian, Karl F. Gauss, a German, and Johann Bolyai, a Hungarian. To explain the details of their discoveries would be beyond the scope of this text, but it should at least be mentioned that they showed that if one were to assume the existence of more than one parallel line through the point off a given line, a valid, yet different, set of theorems could be developed. These theorems are not idle creations of a mathematician's mind but do have some real applications in the physical world.

There is one remaining theorem to be given in this section. Before it is presented, several terms need to be introduced.

DEFINITION 8.2

A line is a **transversal** of two coplanar lines, m_1 and m_2, if it intersects those two lines in exactly two points.

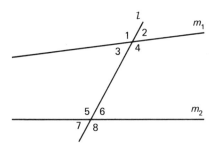

Figure 8-8

If we refer to Fig. 8-8, line l is shown as a transversal. Note that the two lines, m_1 and m_2, do not have to be parallel for l to be a transversal. These lines meet to form angles, some of which, in pairs, have special names because of their positions relative to each other. If two angles are "between" the lines m_1 and m_2 and their interiors are on different sides of the transversal, they are called a **pair of alternate interior angles.** One such pair is made up of $\angle 3$ and $\angle 6$; the other pair is $\angle 4$ and $\angle 5$. The pair made up of $\angle 1$ and $\angle 8$ is called a **pair of alternate exterior angles.** What is the other pair of alternate exterior angles? Finally, if two angles are in the same relative position, they are called corresponding angles. In Fig. 8-8, the pairs of **corresponding angles** are 1 and 5, 2 and 6, 3 and 7, and 4 and 8.

The following theorem tells what is true if, essentially, one of these pairs is made up of congruent angles. The theorem is stated in terms of a pair of alternate interior angles; the other pairs are mentioned in the several corollaries to the theorem. The statements of these corollaries will be given —their proofs will be left to the exercises.

THEOREM 8.8

If two lines are cut by a transversal so that a pair of alternate interior angles are congruent, then the lines are parallel.

Before we give the proof of this theorem, several comments should be made. First, it should be noted that if one pair of alternate interior angles are congruent to each other, then the other pair are congruent also. This fact follows immediately from the fact that the members of one pair are supplements of the other pair, and so, if the first pair is congruent, then so must the second be congruent. Thus, in the proof that follows we choose freely either of the two pairs of alternate interior angles for the proof.

The other remark is that Theorem 8.8 is a generalization of Theorem 8.5. Recall that Theorem 8.5 hypothesized the truth of the fact that a pair of angles, which we now know to be alternate interior angles, were right angles, and thus congruent. Here we demand only congruence and not any specific measure. It is interesting to note that the proof of Theorem 8.8 is almost identical to that of Theorem 8.5. Perhaps the question of interest at this point is whether the facts of Theorem 8.5 are needed for the proof of Theorem 8.8? In other words, in the logical chain of statements that we are slowly building up, is it possible to prove Theorem 8.8 without the aid of Theorem 8.5?

Proof of Theorem 8.8: Refer to Fig. 8-9. The proof is an indirect one, which we begin by assuming $\overleftrightarrow{AB}$ is not parallel to $\overleftrightarrow{CD}$. Remember that we hypothesize the truth of the statement $\angle ABC \cong \angle BCF$. (Of course, the picture does not seem to show this fact, but the picture is drawn this way to show the assumption of the meeting of $\overleftrightarrow{AB}$ and $\overleftrightarrow{CD}$.   Remember that we wish to show that this intersection is not the true situation.)

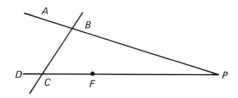

Figure 8-9

If $\overleftrightarrow{AB}$ and $\overleftrightarrow{CD}$ do not meet at some point P, as shown in Fig. 8-9, then $\triangle BCP$ is formed.   The Exterior Angle Theorem in this situation guarantees that $m(\angle ABC) > m(\angle BCF)$.   This fact contradicts the assumed fact, $\angle ABC \cong \angle BCF$ [or, what is equivalent, $m(\angle ABC) = m(\angle BCF)$]. This contradiction is the one we seek in our indirect proof strategy; therefore, we may conclude that the statement "$\overleftrightarrow{AB}$ is not parallel to $\overleftrightarrow{CD}$" is false. Our study of logic assures us that this statement is equivalent to saying $\overleftrightarrow{AB} \parallel \overleftrightarrow{CD}$, which is what is to be proven.

Here are the corollaries to this theorem.   Their proofs, if one uses Theorem 8.8, are short, *direct* proofs.

COROLLARY 8.9

If two lines are cut by a transversal so that a pair of corresponding angles are congruent, then the lines are parallel.

COROLLARY 8.10

If two lines are cut by a transversal so that a pair of alternate exterior angles are congruent, then the lines are parallel.

COROLLARY 8.11

If two lines are cut by a transversal so that a pair of interior angles on the same side of a transversal are supplementary, then the lines are parallel.

One last word of caution.   This last theorem and its corollaries do not depend upon the Parallel Postulate for their truth.   The Parallel Postulate was stated when it was for pedagogical reasons and not because it was logically necessary for Theorem 8.8.

EXERCISE SET 8.2

1. Prove Corollary 8.9.

2. Prove Corollary 8.10.

3. Prove Corollary 8.11.

4. This exercise uses definitions particular only to this problem. A *c*-line is a line that passes through the center of a sphere. An *h*-line is a line that is perpendicular to some *c*-line.
   (a) Can two *c*-lines be parallel?
   (b) Can two *h*-lines be parallel?
   (c) Can two *c*-lines be perpendicular?
   (d) Can two *h*-lines be perpendicular?
   (e) Can a *c*-line be an *h*-line?
   (f) Can an *h*-line be a *c*-line?
   (g) Can a *c*-line be parallel to an *h*-line?
   (h) Would every line be a *c*-line?
   (i) Would every line be an *h*-line?

5. Does the truth of Theorem 8.8 depend upon the truth of the statement in Theorem 8.5?

8.4   SOME EASILY PROVEN THEOREMS ON PARALLELS

In this section there will be given some theorems that are a consequence of the Parallel Postulate. Several of them are the converses of the theorems of the previous section. We begin with the basic one.

THEOREM 8.12

If two parallel lines are cut by a transversal, then alternate interior angles are congruent.

Proof: Figure 8-10 shows the two parallel lines, $\overleftrightarrow{AB}$ and $\overleftrightarrow{CD}$. The task is to prove $\angle ABC \cong \angle BCD$.

On the *A*-side of $\overleftrightarrow{BC}$, choose point G so that $m(\angle GBC) = m(\angle BCD)$. Therefore, $\angle GBC \cong \angle BCD$. By Theorem 8.8, $\overleftrightarrow{GB} \parallel \overleftrightarrow{CD}$. But the Parallel Postulate states that there is at most one line on point B that is parallel to $\overleftrightarrow{CD}$. Therefore, $\overleftrightarrow{GB} = \overleftrightarrow{AB}$. (Note the use of the "$=$" sign here.) Also, $\overrightarrow{BG} = \overrightarrow{BA}$, which thus implies $\angle ABC = \angle GBC$. Since $\angle GBC \cong \angle BCD$, we may then conclude that $\angle ABC \cong \angle BCD$, which is what was to be proven.

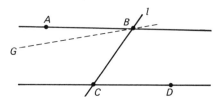

Figure 8-10

This theorem has its corollaries, which turn out to be the converses of the corollaries to Theorem 8.8.

COROLLARY 8.13

If two parallel lines are cut by a transversal, then corresponding angles are congruent.

COROLLARY 8.14

If two parallel lines are cut by a transversal, then alternate exterior angles are congruent.

COROLLARY 8.15

If two parallel lines are cut by a transversal, then interior angles on the same side of the transversal are supplementary.

COROLLARY 8.16

In a plane, if a line is perpendicular to one of two parallel lines, then it is perpendicular to the other.

The proofs of these are left to the exercises.

The next theorem is an immediate result of the Parallel Postulate.

THEOREM 8.17

If two distinct lines are parallel to a third line, then they are parallel to each other.

Proof: Let $m_1 \parallel l$, and $m_2 \parallel l$. We must show $m_1 \parallel m_2$. Although Theorem 8.8 and its corollaries provide means of proving lines parallel, this proof will employ the definition of parallel lines. Using the indirect proof strategy, suppose m_1 and m_2 interest at some point P. If they did, there would be two lines on P parallel to l; this fact contradicts the Parallel Postulate. Therefore, m_1 and m_2 do not intersect. However, to use the definition of parallel lines correctly, it must also be determined that m_1 and m_2 are co-planar. An indirect proof would show that such is the case. The reader is asked to supply the details for himself.

A reading of Theorem 8.17 would seem to indicate that there is a certain transitivity property belonging to the parallel relationship. Indeed, there is, although it is not the strict transitivity that we have used so many times before in this text. For a relation to be truly transitive (aRb, bRc $\Rightarrow aRc$), it is necessary that this relation hold for *all* possible pairs of the elements of the set in question. In the case of the parallel relation, we must exclude the possibility of $a = c$ (in the statement of Theorem 8.17, it would be $m_1 = m_2$), for, if we did not, we would end up with a line being parallel to itself. Thus, the parallel relationship is often described as having a "weak" transitivity property.

We now are at the point where we can prove the theorem that has been used several times. It is the classic result that deals with the sum of the measures of the angles of a triangle.

THEOREM 8.18

The sum of the measures of the angles of a triangle is 180.

Proof: Refer to Fig. 8-11, which pictures $\triangle ABC$, whose angles are marked by the letters x, y, and z. The Parallel Postulate guarantees the existence of a line l on A such that $l \parallel \overleftrightarrow{BC}$.

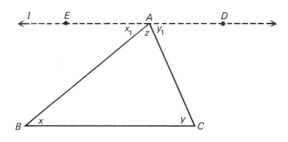

Figure 8-11

By Theorem 8.12:

$$\angle x \cong \angle x_1 \quad \text{and} \quad \angle y \cong \angle y_1$$

which, in terms of measures, state:

$$m(\angle x) = m(\angle x_1) \quad \text{and} \quad m(\angle y) = m(\angle y_1) \qquad (8\text{-}1)$$

Line l is straight and thus may be considered as a straight angle. Using the Angle-Addition Postulate:

$$m(\angle x_1) + m(\angle BAD) = m(\angle EAD) = 180 \qquad (8\text{-}2)$$

Using the Angle-Addition Postulate again:

$$m(\angle z) + m(\angle y_1) = m(\angle BAD) \qquad (8\text{-}3)$$

Substituting from Eq. (8-3) in Eq. (8-2) gives:

$$m(\angle x_1) + m(\angle z) + m(\angle y_1) = 180 \qquad (8\text{-}4)$$

Now, using the equations of Eq. (8-1) in Eq. (8-4) gives the desired result:

$$m(\angle x) + m(\angle z) + m(\angle y) = 180$$

This key theorem also has some immediate corollaries that should be noted. One has already been referred to as Theorem 6.5; it will be restated at this point.

COROLLARY 8.19 (Third Angle Theorem)

If two angles of one triangle are congruent respectively to two angles of a second triangle, then the third angles of each are congruent to each other.

COROLLARY 8.20

The acute angles of a right triangle are complementary.

COROLLARY 8.21

The measure of the exterior angle of any triangle is equal to the sum of the measures of the two remote interior angles.

EXERCISE SET 8.3

1. Prove Corollary 8.13.

2. Prove Corollary 8.14.

3. Prove Corollary 8.15.

4. Prove Corollary 8.16.

5. Since Theorems 8.8 and 8.12 are converses, give the wording of the "if and only if" statement that combines these two.
   Do the same for Corollaries 8.9 and 8.13; for Corollaries 8.10 and 8.14; for Corollaries 8.11 and 8.15.

6. Prove Corollary 8.19.

7. Prove Corollary 8.20.

8. Prove Corollary 8.21.

9. Is it necessary for the three lines of Theorem 8.17 to be coplanar? Explain your answer.

10. Is there a converse to Theorem 8.17?   If so, can you prove it?

11. What is the sum of the measures of the angles of a:
    (a) Quadrilateral
    (b) Pentagon
    (c) Hexagon
    (d) Octagon
    (e) Decagon
    (f) n-gon

12. What is the measure of each angle of a regular:
    (a) Triangle
    (b) Quadrilateral
    (c) Pentagon
    (d) Hexagon
    (e) Octagon
    (f) Decagon
    (g) n-gon

8.5   THE FAMILY OF PARALLELOGRAMS

We now turn our attention to a special subset of the set of quadrilaterals, namely the parallelograms.   The logical place to begin is with the definition of a parallelogram.

DEFINITION 8.3

A **parallelogram** is a quadrilateral with both pairs of opposite sides parallel.

The symbol for parallelogram is $\square$. When a parallelogram is named, the letters of the vertices are listed in consecutive order. Thus, the parallelogram of Fig. 8-12 would be denoted by $\square\ ABCD$, or perhaps, $\square\ ADCB$. Such a notation would then imply, from the definition, the information $\overline{AB} \parallel \overline{CD}$ and $\overline{AD} \parallel \overline{BC}$.

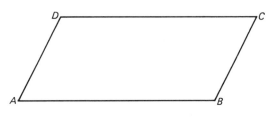

Figure 8-12

Closely related to the idea of a parallelogram is that of a trapezoid. It is essentially defined as a quadrilateral with one pair of opposite sides parallel. Strangely enough, there is lack of agreement among mathematicians as to whether to allow the possibility of the other pair of sides being parallel as well. The arguments are strong for both points of view. Quite arbitrarily, the author has chosen the following definition, which does allow the possibility of both pairs of sides being parallel; thus, the following definition means that parallelograms are a subset of the set of trapezoids. This situation is much akin to allowing the set of equilateral triangles to be a subset of the set of isosceles triangles.

DEFINITION 8.4

A **trapezoid** is a quadrilateral with at least one pair of opposite sides parallel.

The sides that are parallel are called the **bases** of a trapezoid.

DEFINITION 8.5

An **isosceles trapezoid** is a trapezoid that has one pair of opposite sides nonparallel, these two sides being congruent to each other.

As you might expect, the parallelogram has certain properties not generally possessed by a quadrilateral. Many of these properties follow quite directly from a basic theorem.

THEOREM 8.22

A diagonal of a parallelogram, together with the sides of the parallelogram, forms two congruent triangles.

Proof: The truth of the above statement can be seen almost immediately upon inspection of Fig. 8-13. The identically marked angles are congruent, because they are alternate interior angles of the parallel sides—the single marked angles from $\overline{AB} \parallel \overline{CD}$ and the double marked angles from $\overline{AD} \parallel \overline{BC}$. The congruence results from the ASA Postulate.

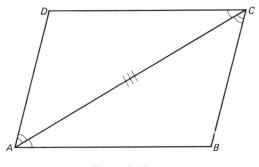

Figure 8-13

From this result and the definition of congruent triangles, we can immediately prove the following corollaries.

COROLLARY 8.23

Any two opposite angles of a parallelogram are congruent.

COROLLARY 8.24

The opposite sides of a parallelogram are congruent.

This second corollary leads to a discussion of an often-heard "definition" of parallel lines as lines that are always the same distance apart. If we wish to measure the distance from a point to a line, this distance is measured along a perpendicular from the point to the line. The distance from a line m_1 to a line m_2 would be from some arbitrary point on m_1. If m_1 and m_2 intersected, then this distance would vary depending upon the selection of the point on m_1. What persons mean when they say that parallel lines are always the same distance apart is that this distance does not vary and is free of the choice of point on m_1.

It might be well to show that this "definition" is not really a definition
at all, but rather is a *property* that follows from what has thus far been
demonstrated in these last few sections.   Figure 8-14 shows two parallel
lines, m_1 and m_2, and two arbitrarily chosen points P and Q on m_1.   From
P and Q are drawn perpendicular lines l_1 and l_2 to m_2 intersecting m_2 at the
points T and V, respectively.   By Theorem 8.5, lines l_1 and l_2 are parallel.
Since we are given that $m_1 \parallel m_2$, we have $PQVT$ a parallelogram.   Corollary
8.24 then tells us that $\overline{PT} \cong \overline{QV}$, which is equivalent to saying m_1 and m_2
are the same distance apart.

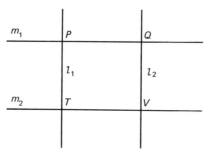

Figure 8-14

Let us examine several more examples of certain "definitions" that are
not really definitions but rather derived properties.   Remember that what
is required of a definition (among other things) is that it give some property
of the object being defined that distinguishes it from the other members of
the class to which it belongs.

As a first example, consider the rectangle, which is often defined as a
parallelogram with four right angles.   The point is: saying it has four right
angles is "over-defining" the rectangle; all that it is necessary to say is that
it has one right angle.   The fact that it indeed does have four can easily be
proven by means of Corollaries 8.15 and 8.23.   A formal definition follows.

DEFINITION 8.6

A **rectangle** is a parallelogram with one right angle.

Two adjacent sides of a rectangle are usually given the special names
of **base** and **altitude**.   Any one of the four sides may be called the base;
when one is called the base, then the side perpendicular to it is called
the altitude relative to that base.

The other example is very similar and deals with an analogous figure,
the rhombus, which is often defined as a parallelogram with four congruent
sides.   Again, this definition is too much.   All that is needed is stated in
the following definition.

DEFINITION 8.7

A **rhombus** is a parallelogram with two adjacent sides congruent.

The congruence of all four sides is established by Corollary 8.24 and the transitivity of segment congruence.

As you no doubt already know, the set of squares is the intersection of the set of rectangles with the set of rhombuses. The definition of a square will be left for you to formulate.

There are many more properties and statements about parallelograms, rectangles, and rhombuses that have not been included in this section. Sometimes the seemingly endless repetition of such statements can lead a student to boredom and/or annoyance. However, these statements are facts that should be encountered by a student of geometry and so have been included in the exercises. Even if you choose not to do the exercises, most of which will be proofs of these statements, you should not pass them over without even a reading.

EXERCISE SET 8.4

1. Define a square.

2. Draw a Venn diagram illustrating the subset relationships that exist among quadrilaterals, parallelograms, trapezoids, rectangles, rhombuses, and squares.

3. Prove: The diagonals of a parallelogram bisect each other.

4. Prove: The diagonals of a rectangle are congruent.

5. Prove: The diagonals of a rhombus are perpendicular to each other.

6. Prove: If the opposite sides of a quadrilateral are congruent, then the quadrilateral is a parallelogram.

7. Prove: If a pair of opposite sides of a quadrilateral are both parallel and congruent, then it is a parallelogram.

8. Prove: Any two consecutive angles of a parallelogram are supplementary to each other.

9. Prove: If the diagonals of a quadrilateral bisect each other, then the figure is a parallelogram.
   (*Hint:* This statement is the converse of the statement in Exercise 3.)

10. The figure below depicts an isosceles trapezoid. Angles *A* and *B* are called the lower base angles of an isosceles trapezoid. Prove that they are congruent to each other.

   (*Hint:* Consider drawing construction lines.)

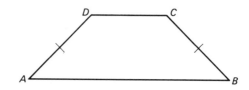

SUGGESTED READINGS

   [9]   Ch. 6

   [11] Ch. 4

   [13] Ch. 10, 11

   [20] Ch. 3

   [22] Ch. 9

   [23] Ch. 4

9

AREA OF
PLANE FIGURES

In this chapter, we shall examine the familiar notion of area. The sets of points that are of interest are polygonal regions; the **area of a region** is a measure assigned to these sets. The material to be presented explains how the assignment of the measure is made.

9.1 THE BASIC CONCEPT OF AREA

Because area of a plane region is a measure, the study of area may be considered within the framework of the familiar ideas of measurement. The measure of a segment was obtained with reference to a segment determined by a unit pair of points; the measure of an angle was determined by reference to the basic angle, the angle of one degree. In an analogous fashion, the measure of a polygonal region must be obtained by reference to some basic polygonal region. Because of the ease of development of the concept of area from it, the square region is chosen as the basic unit.

Before we develop the more commonly used formulas for area, some basic groundwork must be introduced. The theorems and definitions and basic notions developed in the earlier chapters are not enough to allow us to proceed with the development of the area concept. There remain certain properties of area that must be postulated.

POSTULATE 9.1

The area of a polygonal region is a positive number.

This is a reasonable statement when we recall that the measures of both segments and angles were also positive numbers.

POSTULATE 9.2

If two polygons are congruent, then the polygonal regions determined by these polygons have equal areas.

POSTULATE 9.3 (Area Addition Postulate)

If the intersection of two polygonal regions consists of no more than points of their defining polygons, then the area of the union of those two regions is the sum of their areas.

This postulate can be amplified by giving pictorial examples. In Fig. 9-1(a), polygonal region $ABCDF$ is pictured as the union of regions R_1 and R_2, which have only sides $\overline{BG}$ and $\overline{GE}$ in common. Thus, the area of region $ABCDF$ equals the sum of the areas of regions R_1 and R_2. On the other hand, Fig. 9-1(b) shows a region $ABCDF$ that is the union of regions R_1 and R_2, which have the checkered region in common. Here the area of region $ABCDF$ does not equal the sum of the areas of regions R_1 and R_2.

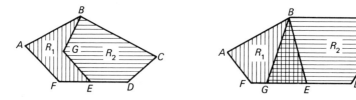

Figure 9-1

The next postulate tells what the area of the basic unit of measurement for area is.

POSTULATE 9.4

The area of a square region having a side with measure 1 is equal to 1.

On the basis of these four postulates, it is possible to develop all the desired formulas.   The one formula that is most important and from which all the others proceed rather directly is the formula for the area of a rectangular region.   Since the development of this formula can be a bit tedious, the formula for the area of a rectangular region shall be postulated also.

POSTULATE 9.5

The area of a rectangular region is equal to the product of the measures of the base and altitude of the associated rectangle, provided these measures have been obtained relative to the same unit of linear measure.

This postulate is often symbolized by $A = bh$, where b stands for the measure of the base, h stands for the measure of the altitude, and A stands for area.   Often, the terms "length" and "width" are used for the two adjacent sides instead of "base" and "altitude."   Then the formula becomes $A = lw$.

In practice, the unit of linear measure referred to in Postulate 9.5 is the length of the side of the square region that is used as a basic unit of measure for area.

EXAMPLES

(a) Rectangle $ABCD$ has base $\overline{AB}$ with a measurement of 6 ft and altitude $\overline{BC}$ with a measurement of $3\frac{1}{2}$ ft.   Thus, rectangular region $ABCD$ has a measure of $6 \cdot 3\frac{1}{2} = 21$.   The measurement of its area is 21 sq ft.

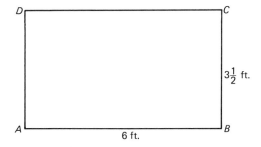

(b) Rectangle *MNPQ* has base $\overline{MN}$ with a measurement of 12 mm and altitude with a measurement of 0.5 cm. The measurements do not have the same unit of measure, and so one of them must be converted. If 12 mm is converted to 1.2 cm, then the measurement of the area of *MNPQ* is 0.60 sq cm; if the 0.5 cm is converted to 5 mm, then the measurement of the area is 60 sq mm.

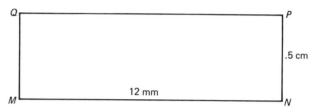

9.2 AREA OF TRIANGLES AND QUADRILATERALS

In this section, the theorems for the area of regions associated with three- and four-sided polygons will be presented. Some of the proofs will be briefly outlined and others will be asked for in the exercises. In each statement, it is assumed that the measures in question have been obtained relative to the same unit of measure.

In order to make the language in the following paragraphs more simple, the convention will be made to use the name of the polygon rather than the polygonal region. For example, we shall say "area of a rectangle" rather than "area of a rectangular region." Of course, it is essential to remember that area is a measure assigned to a polygonal region and not to its defining polygon. The aim here is to seek efficiency and economy of communication, not sloppiness in mathematical expression.

In addition, the symbol "$a(\quad)$" will be used to denote area. In the parentheses, there will be inserted the capital letters naming the vertices of the polygon. Thus, $a(MNPQ)$ will stand for the area of polygon *MNPQ*. However, in certain well-known formulas, a is replaced by A where there is no confusion in thinking A might be the name of the vertex of some polygon.

THEOREM 9.6

The area of a right triangle equals one-half the product of the measures of the two legs of the triangle.

Proof: Refer to Fig. 9-2.   Given $\triangle MNP$ with right angle at N. Q is a point such that $MNPQ$ is a rectangle. $\overline{MP}$ is the diagonal of $MNPQ$ and, by Theorem 8.22, $\triangle MNP \cong \triangle PQM$.   Postulate 9.2 implies that the areas of these two triangles are equal.   Postulate 9.3, the Area Addition Postulate, implies

$$a(MNPQ) = a(MNP) + a(PQM)$$

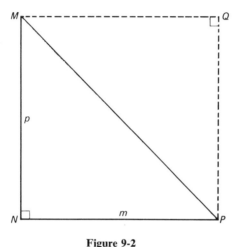

Figure 9-2

Now, Postulate 9.5 states that the area of the rectangle is mp, and so $a(MNP)$ is $\frac{1}{2}\,mp$, which was to be proven.

THEOREM 9.7

The area of a triangle equals one-half the product of the measures of any side of the triangle and the altitude to that side.

(This result is usually symbolized by the formula: $A = \frac{1}{2}bh$.)

Proof: There are three cases to consider: (a) the altitude meets the side at a point that is an endpoint of the side; (b) the altitude meets the side at a point between the endpoints of the side; (c) the altitude does not meet the side at one of its points but rather at a point that lies on the "extension" of the side.   The three cases are pictured in parts (a), (b), and (c), respectively, of Fig. 9-3.

   (a) Essentially Theorem 9.6.
   (b) D is the foot of the altitude.

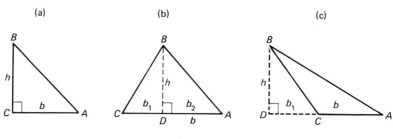

Figure 9-3

$$a(ABC) = a(BCD) + a(BAD)$$

by the Area Addition Postulate.

$$a(BCD) = \tfrac{1}{2}b_1h$$

by Theorem 9.6.

$$a(BAD) = \tfrac{1}{2}b_2h$$

by Theorem 9.6. **Thus,**

$$a(ABC) = \tfrac{1}{2}b_1h + \tfrac{1}{2}b_2h = \tfrac{1}{2}(b_1 + b_2)\,h$$

But, $b_1 + b_2 = b$ by the Segment Addition Postulate.   Therefore,

$$a(ABC) = \tfrac{1}{2}bh$$

which was to be proven.

(c) Proof is left to the reader.

THEOREM 9.8

The area of a square the measure of whose side is s is s^2.

(This result is symbolized by the formula: $A = s^2$.)

Proof: Follows immediately from Postulate 9.5.

THEOREM 9.9

The area of a parallelogram equals the product of the measure of any side and the altitude to that side.

Proof: Refer to Fig. 9-4.   Let $\overline{AB}$ be the base and $\overline{DE}$ be the altitude to that side.

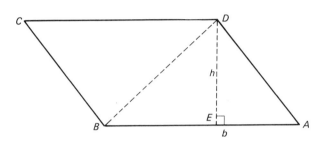

Figure 9-4

$$a(ABCD) = a(ABD) + a(BCD) \qquad \text{(Why?)}$$

$$a(ABD) = \tfrac{1}{2}bh \qquad \text{(Why?)}$$

$$\triangle ABD \cong \triangle CDB \qquad \text{(Why?)}$$

Thus,

$$a(CDB) = \tfrac{1}{2}bh \qquad \text{(Why?)}$$

Therefore,

$$a(ABCD) = \tfrac{1}{2}bh + \tfrac{1}{2}bh = bh \qquad \text{(Why?)}$$

which was to be proven.

THEOREM 9.10

The area of a trapezoid equals one-half the product of the measure of the altitude between the parallel sides of the trapezoid and the sum of the measures of the bases.

Proof: Refer to Fig. 9-5. We are to show that $a = \frac{1}{2}h(b_1 + b_2)$. The proof is left to the reader.

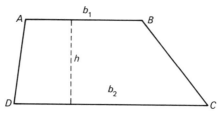

Figure 9-5

EXERCISE SET 9.1

1. Complete the following table of measurements for a rectangular region. Assume all measurements to be exact.

	A	b	h
(a)	—	6 ft	$3\frac{1}{2}$ ft
(b)	—	3.1 cm	1.8 cm
(c)	—	8 in.	$1\frac{1}{4}$ ft
(d)	$3\frac{1}{2}$ sq in.	—	$\frac{3}{4}$ in.

2. Complete the following table of measurements for a triangular region. Assume all measurements to be exact.

	A	b	h
(a)	—	1.7 mm	1.8 mm
(b)	—	$2\frac{2}{3}$ in.	$1\frac{7}{8}$ in.
(c)	—	4 ft 6 in.	3 ft 4 in.
(d)	10.08 sq cm	4.8 cm	—

3. Complete the following table of measurements for a trapezoidal region. Assume all measurements to be exact.

	A	b_1	b_2	h
(a)	—	12 mm	4.2 cm	0.03 m
(b)	—	2 ft	2 ft 8 in.	1 ft 6 in.
(c)	36 sq yd	—	2 yd	6 yd
(d)	105 sq cm	21 cm	14 cm	—

4. Use trigonometry to find the area of parallelogram $ABCD$ if $m(\overline{AB})$ $= 7$, $m(\overline{AD}) = 3$, and $m(\angle A) = 40$; if $m(\overline{AB}) = 13.2$, $m(\overline{AD})$ $= 4.6$, and $m(\angle A) = 45$.

5. Complete the proof of Theorem 9.7.

6. Give the proof of Theorem 9.10.

7. What would the effect be upon the area
   (a) of a square if the measure of the side were doubled?
   (b) of a triangle if the measure of its altitude were halved?
   (c) of a triangle if the measure of its base were tripled?
   (d) of a parallelogram if the measure of its base were doubled?
   (e) of a rectangle if the measure of its altitude were quadrupled?
   (f) of a triangle if the measure of its altitude were doubled and the measure of its base were halved?

8. Prove: The median of a triangle divides the associated triangular region into two triangular regions that have the same area.

9.3 PYTHAGORAS REVISITED

In Section 6.4, the Pythagorean Theorem was stated in algebraic terms. In his treatment of the subject, Euclid did not use algebraic notions but, rather, treated this theorem in strictly geometric terms. Figure 9-6 pictures right triangle ABC with squares constructed upon the three sides. The Pythagorean Theorem states that the area of square $ADEB$ equals the sum of the areas of squares $ACHJ$ and $BCGF$. This section presents two proofs of this theorem using the material of the previous two sections.

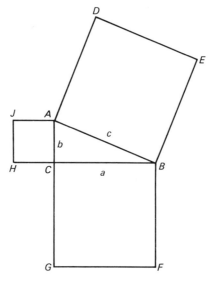

Figure 9-6

For the first proof, we shall refer to Fig. 9-7.   Square *ABCD* is pictured with points *E*, *F*, *G*, and *H* on the sides, so that each side of the square is divided into segments of measure x and y.   The quadrilateral formed by joining points *E*, *F*, *G*, and *H* in succession is a square, the length of whose side is denoted by z.   Can you prove that it is a square?   Using $\triangle AHG$ as a right triangle, we can show that $x^2 + y^2 = z^2$.

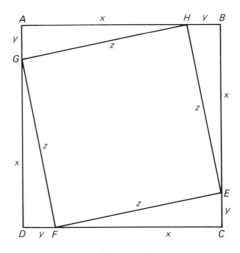

Figure 9-7

Proof: $a(ABCD)$ = Area of four small triangles + $a(HEFG)$.   (Why?)

$$a(ABCD) = (x + y)^2 = x^2 + 2xy + y^2$$

$$\triangle AHG \cong \triangle BEH \cong \triangle CFE \cong \triangle DGF \qquad \text{(Why?)}$$

Thus,

$$a(AHG) = a(BEH) = a(CFE) = a(DGF) = \tfrac{1}{2}xy \qquad \text{(Why?)}$$

$$a(HEFG) = z^2$$

Therefore,

$$x^2 + 2xy + y^2 = 4(\tfrac{1}{2}xy) + z^2$$

This last equation simplifies to the desired $x^2 + y^2 = z^2$.

If the above proof seems somewhat "tainted" by the use of some algebra in its final steps, then the second proof should appeal to the person who wants to see a "pure" geometric proof.   It is a bit more difficult than the above proof but is interesting nevertheless because of its ingenuity.   Figure 9-6 is reproduced as Fig. 9-8 with some auxiliary lines drawn. $\overline{CN}$ is perpendicular to both $\overline{AB}$ and $\overline{DE}$, meeting them at points M and N, respectively.

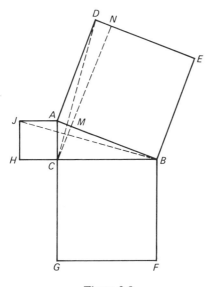

Figure 9-8

Proof: *ADNM* is a rectangle.   (Why?)

$$a(ADNM) = 2 \cdot a(ADC)$$ They have the same base ($\overline{AD}$)
and the same altitude ($\overline{AM}$).

$$a(AJHC) = 2 \cdot a(AJB)$$ They have the same base ($\overline{AJ}$)
and the same altitude ($\overline{AC}$).

The plan is now to show $\triangle ADC \cong \triangle ABJ$ and thus have equal areas. This part of the proof is left to the exercises.

Therefore, it may be concluded that $a(AJHC) = a(ADNM)$.

By means of a similar argument, it may be shown that $a(BCGF) = a(DMNE)$.

The Area Addition Postulate is then used to show that the area of the large square *ABED* = area rectangle *ADNM* + area rectangle *BMNE*. Combining all of this, we have proven the Pythagorean Theorem stated in geometric terms:

THEOREM

The area of a square constructed upon the hypotenuse of a right triangle equals the sum of the areas of the squares constructed upon the two legs of the right triangle.

9.4 AREAS OF POLYGONAL REGIONS

The area of any polygonal region associated with a convex polygon of four or more sides may be found by partitioning the region up into nonoverlapping triangular regions and then finding the area of each triangular region. One method of partitioning would be to choose some vertex of the polygon and then draw all possible diagonals; another would be to choose any point in the interior of the polygon and then connect this point to all vertices of the polygon. Theoretically, finding the area of each triangular region is always possible; practically, it may present some real problems unless sufficient information is known about the parts of the component triangular regions.

In the case of regular polygons, this problem is considerably simplified, since these polygons possess a unique point called the **center**, a point equidistant from all vertices.

Finding the area of a regular polygon will be illustrated by using a regular octagon, as pictured in Fig. 9-9. Each of the eight triangles is an isosceles triangle, and all eight are congruent to each other. The dotted segment, whose measure is denoted by a, is the perpendicular distance from the center of the polygon to the side. This segment is called an **apothem** of the polygon. The area of one triangle is $\frac{1}{2}ba$; thus, the area of a polygon is $8 \cdot (\frac{1}{2}b \cdot a)$. If we use the associative and commutative properties of multiplication, this expression may be written as $A = \frac{1}{2} \cdot (8 \cdot b) \cdot a$. The number, $8b$, is the sum of the measures of the sides of the polygon.

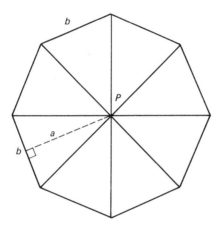

Figure 9-9

In general, the sum of the measures of the sides of any polygon is called the **perimeter** of the polygon. Thus, we may generalize from this example to the general formula for the area of any regular polygon. It is:

$$A = \tfrac{1}{2}ap$$

where a is the measure of the apothem and p is the perimeter.

To continue the example of the regular octagon, suppose the length of one of the sides is 6 in., and we are asked for the area. Since the sum of the measures of the central angles at P is 360, each one is $45°$. Thus, each of the base angles of the small isosceles triangles is $67.5°$. We turn to trigonometry for assistance in finding the length of the apothem. The necessary parts for the solution are pictured in Fig. 9-10. The altitude to the base of an isosceles triangle bisects that base, and we then get the length of the short leg to be 3 in. $a/3 = \tan 67.5° = 2.42$ (from a trigonometric table) Thus, $a = 3 \cdot 2.42 = 7.26$. Now, using the formula above, we get:

$$A = \tfrac{1}{2} \cdot 7.26 \cdot 48 = 174.24 \text{ sq in.}$$

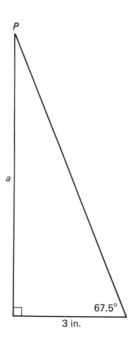

Figure 9-10

EXERCISE SET 9.2

1. Find the perimeter and area of the equilateral triangles whose sides have a measurement of:
   (a) 4 cm   (b) 6 in.   (c) 2.1 mm   (d) $6 \cdot \sqrt{3}$
   (e) s linear units
   (Assume all measurements to be exact.)

2. Find the length of the side of the equilateral triangle that has the same numerical measure for both its perimeter and its area.

3. Find the perimeter and area of the regular pentagons whose sides have a measurement of:
   (a) 8 yd   (b) 0.75 m   (c) s linear units
   (Assume all measurements to be exact.)

4. Find the perimeter and area of the regular hexagons whose sides have a measurement of:
   (a) 9 in.   (b) 1 ft 4 in.   (c) $3\sqrt{2}$
   (d) s linear units

5. In the first proof of Section 9.3, prove that quadrilateral $DEFG$ is a square.

6. In the second proof of Section 9.3, prove that $\triangle ADC \cong \triangle ABJ$.

7. The familiar five-pointed star shown below is often called a penta-gram. What is the sum of the measures of the angles at the "points" of the star?

8. Find the area of the polygonal region pictured below.

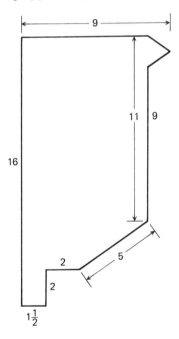

9. Prove that the sum of the perpendicular distances from point P in the interior of an equilateral triangle to the three sides of the triangle is the same number regardless of the placement of point P.

9.5 FINDING THE AREA OF IRREGULARLY SHAPED REGIONS

All the regions whose areas we have been finding have been polygonal regions. As noted earlier, a polygon is a special case of the larger class of objects known as simple closed curves. It is of interest to find the area of any region enclosed by a simple closed curve. To do it properly, we need a knowledge of calculus. However, to get the "feel" of how it is done, intuition is sufficient.

Consider the region enclosed by the simple closed curve C, as pictured in Fig. 9-11. On top of this region is superimposed a **grid** composed of little squares. Some of the little squares are subsets of the region; some intersect the region partially; and some are disjoint with the region. Since the measurement of area is given as the number of square units, we can use the number of square units in the figure to get the area of the region. We count the number of squares completely within the region enclosed by the curve. By the Area Addition Postulate, the sum of the areas of these squares will give the area of the polygonal region that the squares, taken together, compose. The area of this polygonal region is less than or equal to the area of the region. Let us call the number of such square units the **minimum** to be symbolized by *min*. If we let the **maximum** (symbolized by *max*) be the number of square units of the grid such that any part of the square is within the region, then the area of the irregular region will be less than or equal to the maximum. Symbolically, these facts are written:

$$min \leq A \leq max$$

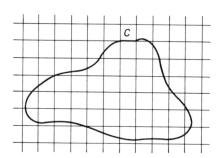

Figure 9-11

where A represents the area of the irregular region.

The minimum and the maximum are only approximations of the area. To get closer approximations, visualize subdividing the side of each original square of the grid into two congruent parts. Thus, each original square is replaced by four smaller squares. If we count the number of smaller squares and then convert back to the original unit of measure, the number of square units in the minimum will either get larger or stay the same. Similarly, the number of square units in the maximum will either get smaller or stay the same. As we repeat this process of making the side of the squares in the grid smaller and smaller, the maximum and the minimum will converge on the same number from above and below—this "same number" turns out to be the area of the irregular region.

9.6 COMPUTING WITH PHYSICAL MEASUREMENTS

In the preceding sections, the given measurements of the theoretical polygons discussed were exact. As we have seen, exact measurements do not exist in the world of real measurements; rather, we deal with approximations. Several examples will be given to point up the caution that should be exercised when computing with such data. At this time, the reader should review the contents of Chapter 4.

EXAMPLES

(a) Find the perimeter of an object rectangular in shape with the dimensions of $3\frac{1}{8}$ in. and $4\frac{5}{8}$ in.

The precision unit is $\frac{1}{8}$ in., so the g.p.e. is $\frac{1}{16}$ in. The first measurement could be between $3\frac{1}{16}$ in. and $3\frac{3}{16}$ in.; the second between $4\frac{9}{16}$ in. and $4\frac{11}{16}$ in. If we take the smallest possible measurement for both, the perimeter is $15\frac{4}{16}$ in.; the largest possible for both yields a perimeter of $15\frac{12}{16}$ in.—a range of $\frac{1}{2}$ inch. Thus, the g.p.e. is $\frac{1}{4}$ in., and the measurement of the perimeter should be reported as $15\frac{1}{2}$ in.

(b) Find the area covered by the rectangular object described in example (a).

If we use the smallest possible measurements, the area is $\frac{3577}{256}$ sq in.; with the largest possible, the area is $\frac{3825}{256}$ sq in.—a difference of $\frac{248}{256}$ sq in., or almost one square unit! If we had been dealing in feet or meters, the difference would have sizeable. The best answer to give is the product of $3\frac{1}{8}$ and $4\frac{5}{8}$, which is $\frac{925}{64}$ sq in., although one should recognize that this number is not the average of the above two measurements of area.

(c) Find the area of a square region whose side has a measurement of 5.6 m.

A measurement of 5.6 m could stand for any measurement between 5.55 m (whose square is 30.8025 sq. m) and 5.65 m (whose square is 31.9225 sq m). To report the measurement of the square as $(5.6)^2$ or 31.36 sq m would be unwise, since such a figure carries the implication that the real measurement lies somewhere between 31.355 sq m and 31.365 sq m, which is not the case. The rule most generally quoted in situations such as this is that the answer can have no more significant digits than either of the factors of the answer. Application of this rule suggests the best answer of "approximately 31 square meters." Rigid and blind adherence to such rules is not advocated however.

EXERCISE SET 9.3

1. Using the examples of the previous section as a guide, report the range of measurements in which the actual measurement of perimeter and area of the following figures lies.
   (a) A square whose side is measured as 1.3 ft.
   (b) A square whose side is measured as 12.46 cm.
   (c) A rectangle whose sides are measured as 4.7 in. and 3.2 in.
   (d) A rectangle whose sides are measured as 0.03 m and 0.13 m.
   (e) An equilateral triangle whose side is measured as 5.5 in.

2. In the discussion regarding the area of an irregular region, is it possible that the square units that are counted as the minimum might not form a polygonal region? Explain your answer.

3. Can the grid system of computing areas of irregular regions be applied to finding the areas of polygonal regions?

SUGGESTED READINGS

[11], Ch. 8

[20], Ch. 4

[22], Ch. 11

[24], Ch. 10

[26], Ch. 8

10

GEOMETRY IN THREE DIMENSIONS

Much of the material presented thus far has been limited to figures lying in the plane. The emphasis now turns to a class of figures that can best be described as **noncoplanar**. A formal, rigorous presentation would require the introduction of many new postulates, definitions, and theorems. Instead, this presentation will return to the more intuitive, informal approach. It will be centered around the three-dimensional analogue of the curve; namely, the surface.

10.1 SURFACES

In two-dimensional geometry, the curve was seen to be the basic figure when one considered geometric figures consisting of more than one point; in three dimensions, the **surface** is the basic figure. In much the same way that the curve was difficult to define precisely, the surface is also a hard term to make precise. Yet, it is intuitively clear to most persons what is meant by a surface. Being a set of fixed points, it does not move; however, it can be described by means of motion of a geometric object. As we thought of a

moving point tracing out a curve by means of moving along some path, it is
also possible to think of a surface as being generated by a curve sweeping
out a set of points in space. In a way analogous to the curve, if the surface
does cross itself, it does so in a set of points that is a curve, and it also has
the property of continuity, which means that it is possible to "move" from
any point of the surface to any other point of the surface along a path that
stays on the surface.

A surface may be finite or infinite in its extent—in either case, it includes
an infinite number of points. It may be a closed surface; if so, it encloses
a region of space, called the interior of the surface, so that a segment
connecting a point of the interior with a point of the exterior must intersect
the surface. As with curves, the surfaces that are of interest are those that
are "well-behaved," in that there is some regularity to them. These surfaces
will be examined in turn. Figure 10-1 depicts some surfaces. If the surface
is infinite in extent, only a portion of it may be represented.

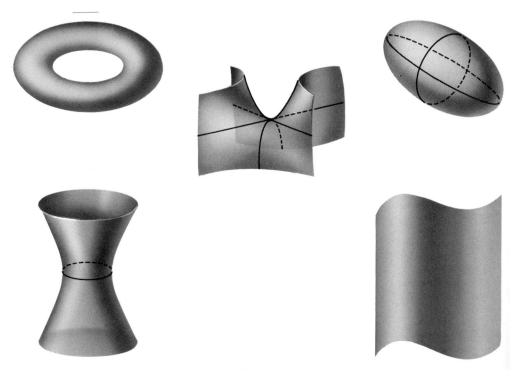

Figure 10-1

Of course, trying to picture a three-dimensional object on a piece of
paper is a most difficult task. Not only is it hard to draw, it is also hard for

many persons to visualize such objects from a two-dimensional drawing. Prospective teachers of space geometry should remember this fact when dealing with young students.

10.2 PLANES

Perhaps the simplest surface is the **plane,** which can be thought of as being generated by a line moving along another intersecting straight line. A plane is sometimes described as the surface that has the property such that if a line intersects it in two points, then the entire line lies in the plane. It is often helpful to think of the plane as being the three-dimensional analogue of the line. As such, many of the relationships of intersection, parallelism, and perpendicularity become clear.

In the three-dimensional world of space geometry, two planes either intersect or they do not. If they do not, then they are said to be **parallel;** if they do, then their intersection is a line. There are no such things as skew planes. The concept of parallel planes is readily extended to three or more planes.

Before we consider what is meant by two planes being perpendicular, the idea of a line being perpendicular to a plane should be introduced first. For a line to be perpendicular to a plane, the line must intersect the plane in a single point and, further, have the property that it is perpendicular to every line in the plane that passes through that point of intersection. This description is illustrated in Fig. 10-2. Only several of the infinitely many lines that lie in plane α and pass through point P are shown.

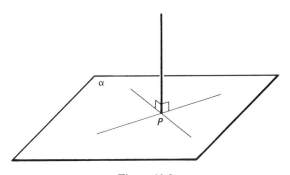

Figure 10-2

To envision perpendicular planes, think of two adjoining walls of a box-shaped room or a wall and a floor. The sides of a box also serve as a good physical model of this relationship. The definition of perpendicular planes may be made using the concept of a line being perpendicular to a plane. It is left to you to formulate this definition.

There are many statements concerning lines and planes, and planes and planes, together with relationships of parallelism and perpendicularity, that are analogues of some of the theorems of previous chapters. Rather than list them and prove them at this point, we shall leave them to the exercises.

EXERCISE SET 10.1

1. Define perpendicular planes.

2. Given below are a number of statements about lines and planes. Use your intuition, reasoning by analogy, and sketches to decide if they are true or false. If they are false, amend or correct them to make them true.

   (a) If a line is parallel to a plane, then it is parallel to every line in that plane.

   (b) If line $\overleftrightarrow{CD}$ is perpendicular to line $\overleftrightarrow{DF}$ of plane β, then it is perpendicular to every line in plane β.

   (c) If line $\overleftrightarrow{CD}$ is perpendicular to plane β, then it is perpendicular to every line in plane β.

   (d) Two planes perpendicular to the same line are parallel.

   (e) If two lines are perpendicular to a third line at the same point, then those two lines are coplanar.

   (f) Through a given point not on a plane, there is exactly one line perpendicular to that plane.

   (g) On a given point not on a plane, there is exactly one plane that is perpendicular to the first plane.

   (h) Two planes perpendicular to a third plane are parallel.

   (i) Two planes parallel to a third plane are parallel.

   (j) Through a given point not on a plane, there is exactly one line parallel to that plane.

   (k) Through a given point not on a plane, there is exactly one plane parallel to that plane.

   (l) Two lines perpendicular to a given plane are parallel.

   (m) Two lines parallel to a given plane are parallel.

   (n) Two skew lines are never coplanar.

3. In the figure on p. 221, $\overline{CD}$ is perpendicular to plane α at point D. $\overline{DX} \cong \overline{DY}$. Prove $\overline{CX} \cong \overline{CY}$.

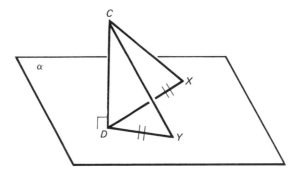

4. Prove that if a plane lies on exactly one of two parallel lines, then it is parallel to the second line.

10.3 DIHEDRAL AND POLYHEDRAL ANGLES

If a plane is the three-dimensional analogue of the line, then a half-plane is the analogue of a half-line. A ray consists of a half-line and its endpoint, and so its counterpart in space is the union of the half-plane and its boundary line. The name for this union is a **closed half-plane.** Our aim is to discuss the geometric figure that is the analogue of an angle—the union of two non-collinear rays with the same endpoint. In the spirit of the previous sentences, we must define such a figure as the union of two noncoplanar closed half-planes sharing the same boundary line. This figure is called a **dihedral angle** and is illustrated in Fig. 10-3. The half-planes are called the **faces** of the dihedral angle and the boundary line is called the **edge** of the angle. There is a standard notation for naming dihedral angles. It is A-BC-D, where A is a point in one of the faces, $\overleftrightarrow{BC}$ is the edge, and D is a point in the other face. A partially open book is a good physical model.

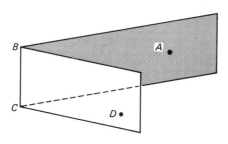

Figure 10-3

It is possible to assign a measure to a dihedral angle by the process of associating a plane angle with the dihedral angle. If a plane is perpendicular to the edge of a dihedral angle, then the intersection of that plane with the dihedral angle is a plane angle. The measure of the plane angle is the measure of the dihedral angle. If two dihedral angles have the same measure, then they are said to be congruent. Figure 10-4 pictures dihedral angle A-BC-D and its associated plane angle, $\angle AED$. The labels, acute, right, and obtuse, are appropriate for use with dihedral angles. If a dihedral angle is right, then its half-planes are perpendicular to each other.

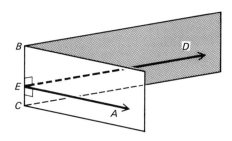

Figure 10-4

If three or more planes intersect at a single point, then we have the "makings" of a figure known as a **polyhedral angle**. Only portions of the planes are included in the figure; since the portion included is not a half-plane, a different explanation is needed. Imagine a polygon and a distinct point P not in the plane of the polygon; then the union of all possible rays $\overrightarrow{PX}$, where X is a point of the polygon, is called a **polyhedral angle**. Point P is called the **vertex** of the angle. If $X_1, X_2, \ldots, X_n$ are the vertices of the polygon, then $\overrightarrow{PX_1}, \overrightarrow{PX_2}, \ldots, \overrightarrow{PX_n}$ are called the **edges** of the polyhedral angle. A plane angle formed by two successive edges is called a **face angle** of the polyhedral angle. The interiors of the face angles are called the **faces** of the polyhedral angle. A polyhedral angle is named P-$X_1X_2 \ldots X_n$.

The polyhedral angle is given a special name, depending upon the number of faces it has. For example, trihedral, tetrahedral, pentahedral, hexahedral, and octahedral are names given to polyhedral angles of three, four, five, six, and eight faces, respectively. The pentahedral angle P-$X_1X_2X_3X_4X_5$ is illustrated in Fig. 10-5. A good physical model of a trihedral angle would be two adjoining walls and a floor of a room.

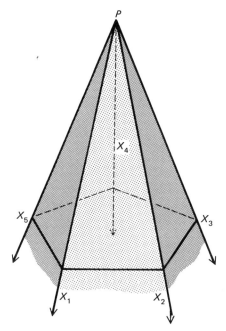

Figure 10-5

EXERCISE SET 10.2

1. Give at least five other physical models of polyhedral angles.

2. Consider dihedral angle A-BC-D. Call the plane determined by A and $\overleftrightarrow{BC}$, γ, and the plane determined by D and $\overleftrightarrow{BC}$, π. What is:

   (a) $\gamma \cap \pi$ (c) $\overleftrightarrow{CD} \cap \gamma$ (e) $\overleftrightarrow{AD} \cap \gamma$ (g) $\overleftrightarrow{AB} \cap \gamma$

   (b) $\overleftrightarrow{AD} \cap \overleftrightarrow{BC}$ (d) $\overleftrightarrow{CD} \cap \pi$ (f) $\overleftrightarrow{AD} \cap \pi$ (h) $\overleftrightarrow{AB} \cap \pi$

3. Would the measure of the plane angle associated with a dihedral angle change if its plane were not perpendicular to the edge of the dihedral angle? If so, would the measure get larger or smaller? Give a plausible reason for your answer.

4. Is there any limit to the measure of a dihedral angle?

5. Is there any limit to the measure of any one of the face angles of a polyhedral angle?

6. Is there any limit to the sum of the measures of the face angles of a polyhedral angle?

10.4 CLOSED SURFACES

There are countless examples of surfaces that are not planes, or unions of planes and subsets of planes. These would be the analogues of the non-straight curves of two dimensions. Some good physical examples are the surface of an inflated balloon, the surface of a football, a saddle used for horseback riding, the paraboloid-shaped reflector of an automobile headlight, an ice cream cone. There are really no bounds in visualizing such surfaces.

The discussion in this section will be restricted to a special class of surfaces; namely, a closed surface, and, in particular, a simple closed surface. Such a geometric figure is best thought of by making reference to the simple closed curve of two dimensions. A simple closed surface partitions space into three distinct and disjoint subsets, the surface itself, and the interior and the exterior of the surface. The interior is a set having the property that *all* segments connecting all possible pairs of points shall be finite in length. Those closed surfaces related to the circle will be studied in the following chapter.

The interior of a simple closed surface need not be convex. The torus, the doughnut-shaped surface, is an example of such a surface; an inflated rubber glove, such as women use for washing dishes, is a physical example of another. However, the discussion will be restricted only to those whose interiors are convex sets. It must be emphasized that the points of the interior are not elements of the set that is the surface.

10.5 POLYHEDRA

If a closed surface consists only of the union of portions of planes, then it is called a **polyhedron** (plural: **polyhedra**). Those portions of planes that make up the surface must, of necessity, be polygonal regions. Various polyhedra are pictured in Fig. 10-6.

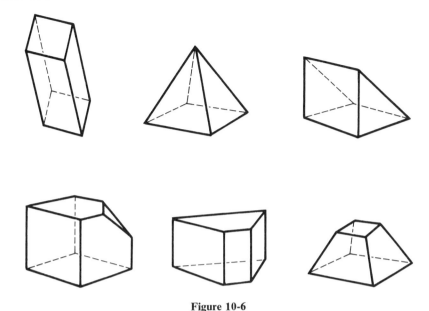

Figure 10-6

The vertex of a polyhedral angle associated with a polyhedron is called a **vertex** of the polyhedron; the intersection of the adjoining polygonal regions is called an **edge**; the polygonal regions are called **faces**.

If certain conditions or relationships exist with regard to the faces and the planes containing those faces, then the polyhedra have special names. The more important of these are the prism and the pyramid, each of which will be covered in more detail.

If all of the faces of a polyhedron are congruent regular polygonal regions, then the polyhedron is said to be **regular**. The best-known example of a regular polyhedron is the **cube**, which has six congruent square regions as faces. There are only five regular polyhedra—the tetrahedron, the hexahedron, the octahedron, the dodecahedron, and the icosahedron, having four, six, eight, twelve, and twenty faces, respectively.

There is a proof of the existence of only five such surfaces that is fairly simple to follow. Although the major details will be left to the exercises, the proof depends upon two main results. One is the answer to Exercise 6 of Exercise Set 10.2, which states that the sum of the measures of the face angles of a polyhedral angle cannot exceed 360. The second is an amazing formula discovered by a famous Swiss mathematician, Leonhard Euler, that states that for any polyhedron, the number of vertices, edges, and faces, denoted by V, E, and F, respectively, are related by the formula:

$$V + F - E = 2$$

EXERCISE SET 10.3

1. Verify Euler's formula for a tetrahedron, a pentahedron (5 faces), a hexahedron, and an octahedron.

   The following sequence of questions is designed to let the reader work out the proof that the only regular polyhedra are the five mentioned in the previous section.

2. What is the measure of an angle of a regular triangle? A regular quadrilateral? A regular pentagon? A regular hexagon?

3. What is the least number of face angles that can make up a polyhedral angle?

4. If the sum of the measures of the face angles of a polyhedral angle must be less than 360, combine the answers of the two previous questions to decide what the largest number of sides is that one of the faces of a regular polyhedron can have.

5. If the face of a regular polyhedron is an equilateral triangle, what are the possibilities for the number of face angles at one vertex?

6. If the face of a regular polyhedron is a square, what are the possibilities for the number of face angles at one vertex?

7. If the face of a regular polyhedron is a regular pentagon, what are the possibilities for the number of face angles at one vertex?

8. Using the results of the previous four exercises, how many possible regular polyhedra are there?

   If the preceding problems have been answered correctly, the reader has shown that there are, as mentioned in the text, only five possible cases for a regular polyhedron. The remaining questions show that all five possibilities do indeed lead to a regular polyhedron.

9. This is the case of a face being an equilateral triangular region and three face angles at each vertex. In this case, and all that follow, let V stand for the number of vertices.

   (a) Under the conditions given, what is the total number of face angles involved in the entire polyhedron?

   (b) Since each face contributes three face angles, the total number of faces can be found by dividing the answer in part (a) by 3. Express this result in terms of the letter V.

   (c) Since each face angle shares an edge with one other face angle, the total number of edges can be found by dividing the answer in part (a) by 2. Express this result in terms of the letter V.

   (d) Substitute the results of parts (b) and (c) into Euler's formula for F and E, respectively. You should get an equation that can be solved for V; solve it.

(e) The answer to part (d) should be a positive integer. Substituting this number into the expressions for F and E, you should get two other positive integers as results. This fact assures you that the possibility given in the conditions of this problem leads to a regular polyhedron. It also tells you which one you get. Which one is it?

10. Repeat steps (a) through (e) of Exercise 9 for each of the other four possibilities:
    (1) An equilateral triangular region as a face with four face angles at a vertex.
    (2) An equilateral triangular region as a face with five face angles at a vertex.
    (3) A square region as a face. (Change part (b) to four face angles instead of three.)
    (4) A regular pentagonal region as a face. (Change part (b) to five face angles instead of three.)

11. Now that you have completed the proof, the five regular polyhedra are pictured in Fig. 10-7.

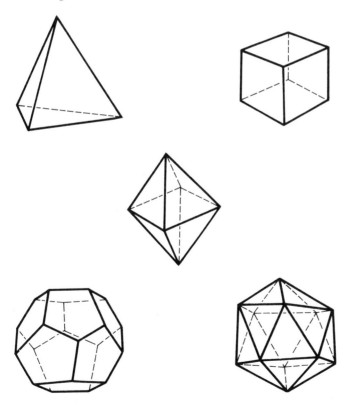

Figure 10-7

10.6 PRISMS

One of the special classes of polyhedra is the set of prisms. Consider two congruent polygons lying in parallel planes. These polygons determine two polygonal regions, called the **bases** of a prism. If corresponding vertices of the polygons are joined by segments to form a set of parallelograms, then the parallelogram regions thus formed are called the **lateral faces** of the prism. The union of the two bases together with the lateral faces is called a **prism.** Various prisms are pictured in Fig. 10-8.

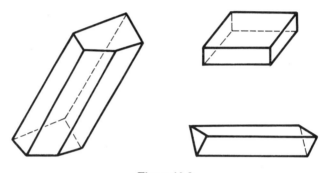

Figure 10-8

If all of the lateral faces are rectangles, then the prism is said to be a **right** prism. Often prisms are named according to the type of polygonal region that the base is. Thus, there are triangular, quadrangular, pentagonal, hexagonal prisms, to illustrate a few. If the bases are parallelograms, then the prism has the special name of **parallelepiped.** Probably the most familiar prism is the one in which the lateral faces and bases are all rectangles—the ordinary box is a good physical model of this prism; the cube is the most specialized example of such a prism.

10.7 PYRAMIDS

Consider a polygonal region lying in some given plane and a fixed point P not in that plane. If line segments are drawn from P to each of the vertices of the polygon, then triangular regions are defined by these segments together with the sides of the polygon. The union of these triangular regions, together with the given polygonal region, constitutes a geometric figure known as a **pyramid.** Several pyramids are illustrated in Fig. 10-9.

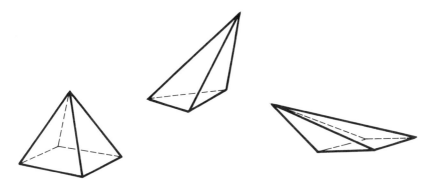

Figure 10-9

The polygonal region is called the **base,** and the pyramid is named according to the type of polygon, in a fashion similar to prisms.   The triangular regions are called **faces** of the pyramid, while the fixed point is called the **apex.** It should be noted that a perpendicular line drawn from the apex to the plane of the base need not intersect that plane in the interior of the base.   Such a perpendicular is called the **altitude** of the pyramid. This situation is shown in Fig. 10-10.   A pyramid is named P-ABC . . ., where P is the apex and A, B, C, etc. are the vertices of the base.

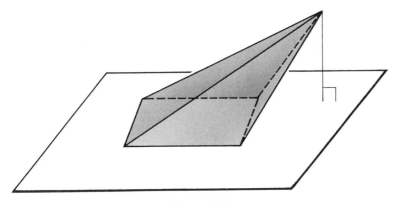

Figure 10-10

If the base is a regular polygonal region and if the perpendicular from the apex to the plane of the base contains the center of the base, then the pyramid is said to be a **regular** pyramid.

EXERCISE SET 10.4

1. Complete the following table:

Prism	No. of Vertices	No. of Edges	Faces
triangular			
quadrangular			
pentagonal			
hexagonal			
octagonal			

2. Verify Euler's formula for each of the above-named prisms.

3. What is another name for a hexahedron that is also a prism?

4. Give an informal proof that the planes containing the lateral faces of a right prism are all perpendicular to the planes containing the bases of that right prism.

5. Describe all the possible intersections of a plane with a regular quadrangular pyramid.

6. Decide if the following statements are true or false. If false, correct them to make them true.

   (a) The bases of a prism are congruent.
   (b) All intersections of a plane with a triangular prism, except the empty set and singleton set possibilities, are triangles that are similar to each other.
   (c) If a prism has square regions for faces, then it is a cube.
   (d) If a prism has square regions for bases, then it is a cube.
   (e) All pyramids are prisms.
   (f) All pyramids are polyhedra.
   (g) The segments connecting the apex of a regular pyramid to the vertices of its base are congruent.
   (h) The lateral faces of a prism are rectangles.
   (i) Some polyhedra are not prisms.
   (j) A tetrahedron is a pyramid.

10.8 SURFACE AREA AND VOLUME

Thus far, the discussion has concerned itself with simple closed surfaces. We now shift attention to the union of these surfaces with their interiors. Such geometric objects are called **solids;** in particular, we shall be interested in the measures assigned to these solids. Because of certain difficulties inherent in determining the measures of all solids, the material in this section will be restricted to particular cases of polyhedra.

The surface area of a solid is the analogue of the perimeter of a simple closed curve. The surface area is the sum of the areas of the bounding regions of the solid. If the solid is a polyhedron, the faces that bound the solid are all polygonal regions, the areas of which are determined by the methods outlined in Chapter 9. Formulas can be developed for some of the simpler polyhedra, such as prisms and pyramids; however, the simple expedient of adding up the areas seems to be sufficient.

The development of formulas for volume follows very closely the development of those for area with addition of the factor for the third dimension. Rather than go through the development in detail, we shall give only the important results.

The measure assigned to solids is determined with reference to a unit of measure, which is also a solid. For the same reason that the square was chosen as the unit for area, the cube is chosen as the basic unit of measure for volume. Thus, the volume of a cubical solid having an edge with measure 1 is equal to 1. If the *measurement* of the edge is 1 ft, then the *measurement* of the volume of the cubical solid is 1 cu ft.

To develop the formulas for solids, one formula is postulated. It is the formula for the rectangular solid, which is the solid associated with the right prism whose bases are rectangles. At this point, it should be noted that many persons use expressions such as "volume of a prism" or "volume of a pyramid." As we have defined these figures, a prism or a pyramid is just the "shell" of the solid and, as such, has no volume. Thus, such expressions are technically incorrect; yet, they do have a desirable economy of words. With this in mind, the remainder of the exposition will use these economical expressions with the understanding that the true meaning is the volume of the solids associated with such surfaces.

POSTULATE 10.1

The volume of a rectangular solid is equal to the product of the measures of the three distinct edges of the solid, provided these measures have been obtained relative to the same unit of linear measure.

This postulate is usually symbolized by the formula $V = lwh$, where l, w, and h are the symbols for the measures of the three distinct edges; it is pictured in Fig. 10-11. The figure shows a case where the three measures are positive integers, although the formula holds for any positive real numbers. The measure of volume of the pictured solid is $V = lwh = 5 \cdot 4 \cdot 2 = 40$.

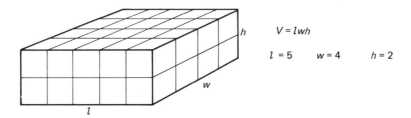

Figure 10-11

If this formula is written, using the associative law for multiplication, it is $V = (lw)h$; the expression (lw) is the expression for the idea of the base of this prism; if we let B stand for this expression, then the formula for volume can be written as $V = Bh$. In this formula, the number h then stands for the *altitude* or *height* of the prism, which is the perpendicular distance between the two parallel planes containing the bases of the prism. Since the two bases are congruent, the symbol B then stands for the area of either base.

This general formula, $V = Bh$, may be extended to prisms other than right rectangular prisms. First, consider a rectangular prism whose lateral faces are merely parallelograms rather than rectangles. If a side view only is pictured, we get the configuration shown in Fig. 10-12. The area of the parallelogram and the rectangle are equal since their dimensions, l and h, are the same. If we consider the third dimension, w, to be the same for both (it cannot be seen in the picture given), it would seem reasonable to hypothesize that the volumes of both prisms, which these pictures represent, should be the same. Consideration of a physical example should convince you that they are, indeed, the same.

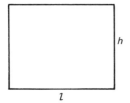 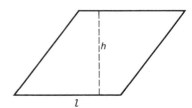

Figure 10-12

Imagine a block of wood or some other substance in the shape of a rectangular solid. If this solid were passed through a slicing machine, very thin wafers in the shape of rectangular regions would be formed. If we assume no loss from the slicing, the volume of the sliced stack would be equal to the volume of the original solid. But, suppose the slices were not stacked so that their vertical edges were perpendicular to the plane of the bottom slice but rather were put in a "slanting" position. The slanting position does not cause any loss of volume; thus, the new solid formed by the slices has the same volume as the original solid. The area of the bases of the two solids is the same, and the heights are the same, whether in the upright rectangular position or in the slanted position. Therefore, the volumes are the same. Perhaps the best, yet most simple way of illustrating this situation, is to take a deck of ordinary playing cards, whose volume is the same whether it is erect or leaning to one side. All of this discussion is pictured in Fig. 10-13.

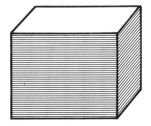 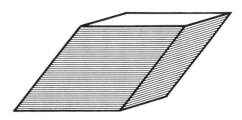

Figure 10-13

The previous paragraphs may be summarized by saying that the volume of a rectangular prism may be found by the formula

$$V = Bh$$

where B is the area of the rectangular base, and h is the measure of the height between bases.

With this formula in mind, it is easy to generalize even further to cases of prisms whose bases are not rectangles. We could shift to a parallelogram base, then to a triangular base, and then to a general polygonal base. In all cases, the result would be the same—the volume may be found by multiplying the area of the base by the measure of the height between the bases. Thus, the general formula for the volume of all prisms is

$$V = Bh$$

This formula may be used to develop the formula for the volume of a pyramid. To do this, a postulate similar to Postulate 9.3 must be formulated. Informally speaking, this is a volume additivity postulate.

We shall use a triangular prism to develop the formula for a triangular pyramid. The plan is to partition the prism into three pyramids, each of which has the same volume and thus show that one pyramid has a volume equal to one-third the volume of the prism. To proceed, we must postulate the fact that if two pyramids have congruent bases and congruent altitudes, then their volumes are equal. (This is analogous to the situation with prisms that has just been discussed.) Reference is made to Fig. 10-14.

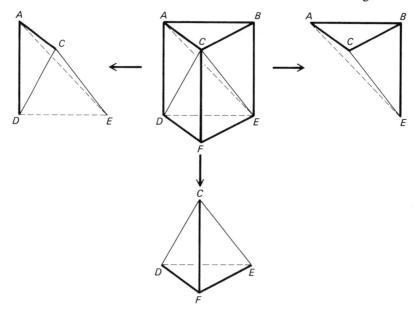

Figure 10-14

Since $ACFD$ is a parallelogram (why?), $\triangle ACD \cong \triangle FDC$. Therefore, pyramids $E\text{-}ACD$ and $E\text{-}FDC$ have equal volumes. (Why are their altitudes congruent?) Because we are dealing with a prism, $\triangle ACB \cong \triangle DFE$. (Why?) Therefore, pyramids $E\text{-}ACB$ and $C\text{-}DFE$ have equal volumes. (Why are their altitudes congruent?) But, pyramids $E\text{-}FDC$ and $C\text{-}DFE$ are the same solid. Therefore, the original triangular prism has been partitioned into three pyramids with equal volumes; thus, we may conclude that the volume of the triangular pyramid is one-third of the volume of the prism.

In summary, the volume of a triangular pyramid may be found by the formula:

$$V = \tfrac{1}{3}Bh$$

where B is the area of its base and h is the measure of its altitude.

By partitioning any polygonal pyramid into triangular pyramids, and then using the above formula repeatedly, and finally applying the Area Addition Postulate with the distributive property, we can develop a formula for the volume of any pyramid.   It is the same as above, where B is the area of the polygonal region that is the base of the pyramid, and h is the measure of the altitude.

EXERCISE SET 10.5

1. Complete the following tables.   Assume all measurements to be exact.
   (a) For a rectangular solid.

V	l	w	h	Total Surface Area
—	4 in.	5.2 in.	2.3 in.	—
—	3.5 ft	2.4 ft	1 ft	—
—	11 cm	1.5 dm	8 cm	—
138 cu in.	6 in.	—	8 in.	—
56 cu ft	—	3 ft 6 in.	9 in.	—

(b) For a triangular prism.

V	B	h
—	18 sq mm	11 mm
—	14.3 sq yd	6.7 yd
144 cu in.	—	9 in.

(c) For a pyramid.

V	B	h
—	13 sq cm	12 cm
96 cu ft	—	12 ft
14.7 cu in.	21 sq in.	—

2. Find the total surface area and volume of a regular tetrahedron whose edge measures 5; if the edge measures 3.

3. Find the total surface area and volume of a cube whose edge measures 0.6; if the edge measures $\sqrt{2}$.

4. The figure below represents a regular square pyramid with altitude $\overline{PQ}$. $\overline{PH}$ represents the altitude of one of the faces (called the **slant height**). Find the volume if $m(\overline{PQ}) = 8$ and $m(\overline{PH}) = 10$. (*Hint:* Use the Pythagorean Theorem.)

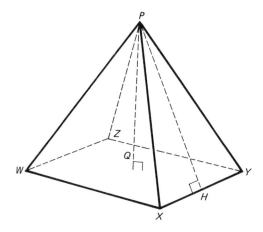

5. Find the volume of the pyramid in Exercise 4 if $B = 100$ and $m(PH) = 13$.

6. Describe a method similar to the grid method of Section 9.5 for finding the area of regions bounded by simple closed curves, which method would find the volume of a solid region enclosed by a simple closed surface.

7. Is there such a thing as symmetry for surfaces and solids? Explain.

8. Verify the formula for volume of a pyramid experimentally by constructing a pyramid and a prism with congruent bases and altitudes. Then, by filling the pyramid with liquid or sand, show that three fillings of the pyramid are needed to fill up the prism.

9. How can six wooden matchsticks be placed to form four congruent equilateral triangles?

10. The figures below show how models of regular polyhedra can be made. By tracing or drawing on heavy cardboard, cutting on the solid lines, folding on the dotted lines and then pasting where necessary, the models can be constructed. Only the configurations for the tetrahedron and the hexahedron are shown. Can you figure out how to make the remaining three regular polyhedra?

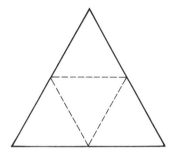 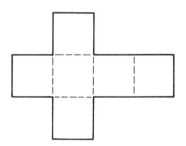

SUGGESTED READINGS

[21] Appendix 1

[25] Ch. 6, 7

[26] Ch. 12

11

CIRCLES, SPHERES, AND RELATED FIGURES

Attention is now directed to circles and spheres, two geometric objects that have perhaps the greatest abundance of physical models in the natural and man-made world. Tires, balls, plates, rims of drinking glasses are but a few of the many examples that help one become familiar with these geometric figures. In geometry, there are many interesting results connected with the study of circles and spheres, making them one of the more fascinating topics to consider.

11.1 CIRCLES AND SPHERES

We proceed directly to some of the basic definitions.

DEFINITION 11.1

Given a point P and a segment $\overline{AB}$, the set of all points Q lying in a plane α such that $\overline{PQ} \cong \overline{AB}$ is known as a **circle**.

DEFINITION 11.2

Given a point P and a segment $\overline{AB}$, the set of all points Q in space such that $\overline{PQ} \cong \overline{AB}$ is known as a **sphere**.

These definitions are somewhat different from definitions appearing in other texts. They were stated in this fashion to point out that they could be made without reference to the word "distance." The point P of both definitions is called the **center** of the circle or sphere and, in neither case, is it a point of the circle or sphere. The measure of the segment $\overline{AB}$ is called the **radius**; the term "radius" is also used to refer to a line segment connecting the center to a point of the circle or sphere. The letter r is generally used to symbolize the radius. If two circles or two spheres have congruent radii, they are said to be congruent.

DEFINITION 11.3

A line segment whose endpoints lie on a circle or sphere is called a **chord** of the circle or sphere.

DEFINITION 11.4

A chord lying on the center of a circle or sphere is called a **diameter** of the circle or sphere.

DEFINITION 11.5

A line intersecting a circle or sphere in two distinct points is called a **secant** of the circle or sphere.

Thus, a chord is a subset of a secant with the property that its interior points lie in the interior of the circle or sphere.

DEFINITION 11.6

A line intersecting a circle or sphere in exactly one point is called a **tangent** to the circle or sphere. In the case of the circle, the tangent must lie in the same plane as the circle.

The single point of intersection in the above definition is called the **point of tangency**.

DEFINITION 11.7

Two circles or spheres that have the same center but different radii are said to be **concentric**.

DEFINITION 11.8

The intersection of a sphere with a plane lying on the center of the sphere is called a **great circle** of the sphere.

DEFINITION 11.9

A plane intersecting a sphere in exactly one point is called a **tangent plane**.

Although many theorems could be proven on the basis of the above definitions, only two will be given here. One will be proven and the other only stated.

THEOREM 11.1

In a circle or a sphere, two chords are congruent if and only if they are equidistant from the center.

Proof: Only the case of the circle will be given; the proof for the sphere is identical. Since the theorem is an "if and only if" statement, this means that it really embodies two statements. These two will be taken in turn. Reference is made to Fig. 11-1.

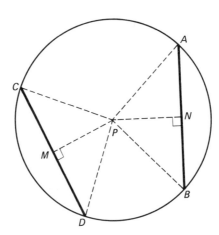

Figure 11-1

(a) "Only if": Assume the truth of $\overline{AB} \cong \overline{CD}$. We are to show that distances measured along perpendiculars are equal; that is, $\overline{PM} \cong \overline{PN}$.

$$\overline{PA} \cong \overline{PB} \cong \overline{PC} \cong \overline{PD}$$

since they are all radii of the same circle. By the hypothesis, $\overline{AB} \cong \overline{CD}$. Therefore, $\triangle PAB \cong \triangle PCD$ by the SSS Postulate. This congruence implies $\angle A \cong \angle C$. $\angle PNA \cong \angle PMC$, because they are both right angles. Therefore, $\angle APN \cong \angle CPM$ by the Third Angle Theorem. Thus, $\triangle PNA \cong \triangle PMC$ by the ASA Postulate; and $\overline{PM} \cong \overline{PN}$ follows directly.

(b) "If": Assume the truth of $\overline{PM} \cong \overline{PN}$ in order to show $\overline{AB} \cong \overline{CD}$. The proof is almost the reverse of the above argument. The difference lies in proving $\triangle PNA \cong \triangle PMC$. The proof of that part involves three corresponding parts that are two sides and a *non*included angle. Ordinarily, such a scheme does not work (refer to Exercise 4 of Exercise Set 5.3); however, it does work when the angle in question is a right angle. The rest of the proof is left to the reader.

THEOREM 11.2

A radius of a circle is perpendicular to a line passing through its endpoint on the circle if and only if the line is a tangent to the circle.

Proof: Omitted.

The above theorem is illustrated in Fig. 11-2. Since the perpendicular is the shortest segment from a point to a line, any other point of the tangent line must lie at a further distance from the center than the point of tangency.

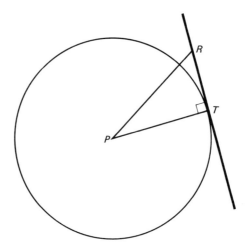

Figure 11-2

Thus, in the figure,

$$m(\overline{PR}) > m(\overline{PT}) = r$$

EXERCISE SET 11.1

1. Complete the proof of part (b) of Theorem 11.1.

2. What is the relationship of the radius of a sphere to a tangent plane when the radius is drawn to the point of tangency?

3. Define the interior of a circle (or sphere) in terms of measure of line segments.

4. Define the exterior of a circle (or sphere) in terms of measure of line segments.

5. Given a point lying in the exterior of a circle, how many tangents to the circle are there lying on that point?

6. Given a point lying in the exterior of a sphere, how many tangents to the sphere are there lying on that point?

7. Given a point lying in the exterior of a sphere, how many tangent planes to the sphere are there lying on that point?

8. Given a line lying in the exterior of a sphere, how many tangent planes to the sphere are there lying on that line?

9. Try to prove your answers to Exercises 5 through 8.

10. Your answer to Exercise 5 should be "two." Prove that these two tangents to a circle from an external point are congruent. (*Hint*: Consult the discussion of part (b) of Theorem 11.1.)

11. There is a rather well-known "brainteaser," which is as follows: If you have three houses in a row, labeled *A*, *B*, and *C*, as shown, and three utility plants in a row, say gas, water, and electricity, can you draw a picture connecting each utility with each house without crossing any lines? Some of the lines are drawn to get you started. Try it before you read the answer below.

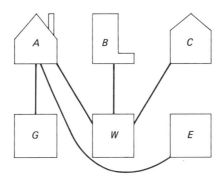

The answer is "no." Could you solve this problem if the pictures of the houses and utilities were located on a surface other than a plane; for example, a sphere?

11.2   CIRCLES AND TRIANGLES

If one were to take a circle and randomly select three points of it, then a triangle would be determined. By our choosing all sorts of positions on the circle, it would seem possible to generate as many differently shaped triangles as could be imagined. Also, if we vary the size of the circle, as many different-sized triangles as desired could be formed. All such triangles are said to be **inscribed** in a circle; all such circles are said to be **circumscribed** about the triangle.

In the foregoing discussion we started with the circles and formed the triangles. If we start with the triangle, an interesting question is raised: Is it possible to circumscribe a circle about *any* given triangle? As one might expect, the answer is in the affirmative. The following paragraphs indicate how this can be done.

The first tool necessary is a theorem regarding a line that is a perpendicular bisector of a segment.

THEOREM 11.3

Any point lying on a perpendicular bisector of a segment is equally distant from the endpoints of that segment.

Proof: If we refer to Fig. 11-3, l is the perpendicular bisector of $\overline{AB}$. We are to show that, for an arbitrarily selected point P,

$\overline{PA} \cong \overline{PB}$

$\overline{PQ} \cong \overline{PQ}$; $\angle PQA \cong \angle PQB$, because they are right angles;

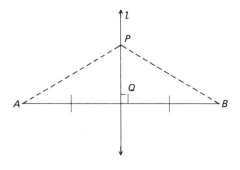

Figure 11-3

$\overline{QA} \cong \overline{QB}$, because Q is the midpoint of $\overline{AB}$. Therefore, the triangles are congruent by SAS, and $\overline{PA} \cong \overline{PB}$ follows at once.

The converse is also true; namely, any point that is equidistant from the endpoints of a segment lies on the perpendicular bisector of that segment. The proof is fairly straightforward and uses the indirect proof technique.

We can now locate the center of the circle that circumscribes any triangle. This point is called the **circumcenter** of the triangle. By methods that will be shown in the following chapter, the perpendicular bisectors of the sides of a triangle may be constructed. Figure 11-4 illustrates an arbitrary triangle, ABC. The perpendicular bisector of $\overline{AB}$ is l, and m is the perpendicular bisector of $\overline{AC}$. P is the point of intersection of l and m. By Theorem 11.3, $\overline{PA} \cong \overline{PB}$ and $\overline{PA} \cong \overline{PC}$. Thus, P is the desired circumcenter, and the circle is drawn with P as center and $\overline{PA}$ as radius.

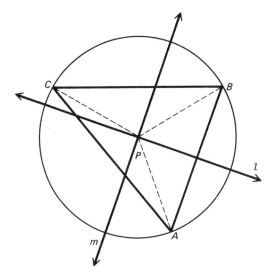

Figure 11-4

Robert Simson, a mathematician of the eighteenth century, discovered an interesting property of the circumcircle. By choosing a point P of the circumcircle that is not a vertex of the triangle and then drawing perpendiculars to the three sides of the triangle, he found that the points of intersection of these perpendiculars with the sides are collinear. Figure 11-5 illustrates this result. $\overline{PX} \perp \overleftrightarrow{AC}$, $\overline{PY} \perp \overleftrightarrow{BC}$ and $\overline{PZ} \perp \overleftrightarrow{AB}$; X, Y, and Z are collinear on line l.

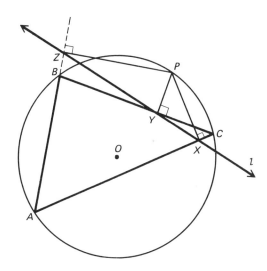

Figure 11-5

There is a similar situation where we can select three random points on a circle and draw tangents to the circle at these points. If two of the three tangents are not parallel to each other, then a triangle will be formed. If we select various triplets of points and various-sized circles, it again seems reasonable to say that all possible triangles would be generated. Such a circle is said to be **inscribed** in a triangle, and its center is called the **in-center** of the triangle.

Again, the reverse question is of interest. Given any arbitrary triangle, is it possible to inscribe a circle? The answer is again affirmative, and, to find the incenter, we need a theorem concerning a property of the bisector of an angle.

THEOREM 11.4

Any point lying on the bisector of an angle is equidistant from the sides of that angle.

Proof: In Fig. 11-6, $\overline{PQ}$ is the bisector of $\angle APB$. We are to show that, for an arbitrarily selected point X on $\overrightarrow{PQ}$, $\overline{XM} \cong \overline{XN}$ when $\overline{XM} \perp \overline{PA}$ and $\overline{XN} \perp \overline{PB}$.

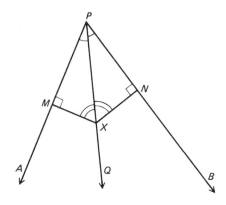

Figure 11-6

The right angles at M and N are congruent because they are right angles; the two small angles at P are congruent, because the larger angle was bisected. Thus, the two angles at X are congruent by the Third Angle Theorem. $\overline{PX} \cong \overline{PX}$. Thus, $\triangle PXM \cong \triangle PXN$ by ASA; and $\overline{XM} \cong \overline{XN}$, the desired result, follows immediately.

The converse of this theorem is also true; namely, if a point is equidistant from the sides of an angle and lies in the interior of the angle, then it lies on the bisector of the angle.

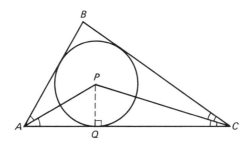

Figure 11-7

The incenter of any triangle is located by bisecting any two of the angles of a triangle. The method of constructing such bisectors is given in the following chapter. From the point of intersection, a perpendicular is dropped to any one of the three sides. The above theorem assures us that all three perpendiculars will be congruent. The length of this perpendicular is the radius of the incircle. This procedure is shown in Fig. 11-7.

EXERCISE SET 11.2

1. In Fig. 11-4, the circumcenter was shown as lying in the interior of $\triangle ABC$. Will the circumcenter of a triangle always lie in the interior of the triangle?

2. In Fig. 11-7, the incenter was shown as lying in the interior of $\triangle ABC$. Will the incenter of a triangle always lie in the interior of the triangle?

3. In Fig. 11-4, only two of the three perpendicular bisectors of the sides of the triangle were shown. To demand that a line be perpendicular to a given segment at the segment's midpoint is sufficient to determine that line uniquely; to further require that it pass through point P, the intersection of two other lines, is to "overdefine" a line, something that should not be done. Thus, we cannot automatically assume that the perpendicular bisector of $\overline{BC}$ in Fig. 11-4 will lie on P—it must be proven that it does, in fact, lie on P. Write out the proof of this fact.

4. In a fashion similar to the situation of Exercise 3, the angle bisector of B in Fig. 11-7 cannot be assumed to lie on P of that figure. It must be proven. Write out the proof of this fact.

5. Prove the converse of Theorem 11.3.

6. Prove the converse of Theorem 11.4.

7. Can a circle be inscribed in or circumscribed about any convex polygon of more than three sides? Can you prove your answer? Is there any special class of polygons that always "works"?

8. Describe what you think may be analogous properties about inscribing and circumscribing spheres with respect to polyhedra.

9. Draw two circles that intersect in two distinct points. The segment connecting these two points is a chord common to both circles. Draw the line connecting the centers of the two circles. What appears to be the relationship between the line joining the centers and the common chord? Can you prove it?

10. If two circles are tangent to the same line at the same point of tangency, the circles are said to be tangent to each other. If the two centers lie in the same half-plane of the tangent line, the circles are said to be internally tangent; if they lie in different half-planes, the circles are said to be externally tangent. What appears to be the relationship between the line joining the centers and the common tangent? Can you prove it?

11. If two coplanar circles are tangent to the same line at two distinct points of tangency, the line is called a common tangent. If the two centers are in the same half-plane of the tangent line, the line is called a common external tangent; if the centers lie in different half-planes of the tangent line, the tangent is called the common internal tangent. Draw all possible configurations of this situation and decide if there are any relationships between these tangents and the line joining the centers.

11.3   CIRCLES, ANGLES, AND ARCS

Certain configurations of circles and angles prove to be of interest because they lead to a way of measuring a circle.

DEFINITION 11.10

If an angle in the same plane as a circle has its vertex at the center of that circle, it is called a **central angle** of the circle.

A central angle of circle P is shown in Fig. 11-8. The only two points the circle and the angle have in common are points A and B of the figure.

That portion of the circle lying in the interior of the central angle, together with points A and B, does have a special name. It is called the arc intercepted by $\angle APB$. But first, let us give a formal definition of arc.

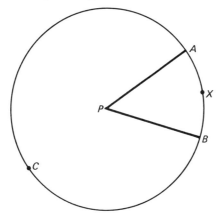

Figure 11-8

DEFINITION 11.11

Given a circle with center P and two distinct points A and B of the circle, an **arc** is the union of A, B, and all points of the circle in the same half-plane of $\overleftrightarrow{AB}$.

If points A and B are the endpoints of a diameter, then the arc is called a **semicircle**. If A and B are not the endpoints of a diameter, the arc is called a **minor arc** if the center of the circle is in the opposite half-plane from the points of the arc; it is called a **major arc** if the center is in the same half-plane with the points of the arc.

The symbol for arc is $\overset{\frown}{AB}$, where A and B are the endpoints of the arc. Such notation may lead to confusion, since A and B are the endpoints of both the major and minor arc. One way to avoid trouble is to choose a third point of the circle "between" the endpoints and write $\overset{\frown}{ACB}$, where the letters A and B still stand for the endpoints and C is the point between the endpoints. Thus, in Fig. 11-8, $\overset{\frown}{ACB}$ would be the symbol for the major arc with endpoints A and B, while $\overset{\frown}{AXB}$ would be the symbol for the minor arc. The other expedient would be to use the words "major" and "minor" as prefixes.

Arcs of a circle may be given a real number as a measure; this number is called **arc degrees** and is symbolized by $m(\overset{\frown}{AB})$. The number assigned to an arc as its degree measure is a function of the measure of the central angle associated with an arc. Because angles have measures less than 180 (as we have agreed upon in this text), the following definition must be made in several parts to take care of the three principal types of arcs.

DEFINITION 11.12

The arc degree measure of:

(1) a minor arc is equal to the measure of its associated central angle.
(2) a semicircle is 180.
(3) a major arc is equal to 360 minus the measure of its corresponding minor arc.

It is important not to confuse angle degrees with arc degrees. Both are alike in that they are nonnegative real numbers, but the former is a measure belonging to angles, whereas the latter is a measure belonging to arcs. The arc degree measure of an arc is independent of the size of the radius of the circle and depends solely upon what portion of the total circle it is. Notice that part (2) of the above definition gives rise to the familiar saying that "there are 360 degrees in a circle."

DEFINITION 11.13

If two arcs of the same or congruent circles have equal measures, then they are said to be **congruent arcs**.

The other angle to be considered here is the angle that has its vertex as a point of the circle. Figure 11-9 illustrates this situation. The angle is said to be inscribed in an arc of the circle.

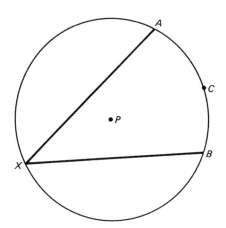

Figure 11-9

DEFINITION 11.14

An angle is **inscribed in an arc** of a circle if the sides of the angle contain the endpoints of the arc and the vertex of the angle is a point of the arc between the endpoints of the arc.

Thus, in Fig. 11-9, $\angle AXB$ is inscribed in arc $\overset{\frown}{AXB}$. The arc $\overset{\frown}{ACB}$ is called the arc **intercepted** by $\angle AXB$. The measure of the inscribed angle is related to the measure of the intercepted arc, as the next theorem shows.

THEOREM 11.5

The measure of an inscribed angle is equal to one-half the measure of its intercepted arc.

Proof: There are three cases to consider: (a) the center of the circle lies on a side of the angle; (b) the center lies in the interior of the angle; (c) the center lies in the exterior of the angle. Figure 11-10 illustrates these three cases. Only the first case will be proven in detail.

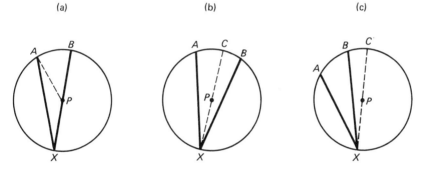

Figure 11-10

(a) Drawing radius $\overline{PA}$ forms $\triangle PAX$ with exterior angle, $\angle BPA$.
$m(\angle BPA) = m(\angle PAX) + m(\angle PXA)$. (Why?) $\overline{PA} \cong \overline{PX}$, because they are radii of the same circle; thus, $\angle PAX \cong \angle PXA$, by the Isosceles Triangle Theorem. Therefore, $m(\angle PAX) = m(\angle PXA)$. (Why?) Using algebra and substitution, we get $m(\angle PXA) = \frac{1}{2} m(\angle BPA)$. But, Definition 11.12 states that $m(\angle BPA)$ equals the measure of minor arc $\overset{\frown}{AB}$. (How can one be sure $\overset{\frown}{AB}$ is always a minor arc?) Again, substitution leads to the desired result that $m(\angle PXA) = \frac{1}{2} m(\overset{\frown}{AB})$.

(b) Draw diameter $\overline{XC}$. Then use case (a) on $\angle AXC$ and $\angle BXC$ and then add. Note that it is necessary to say something to the effect that $m(\overset{\frown}{ACB}) = m(\overset{\frown}{AC}) + m(\overset{\frown}{CB})$. Such a result could either be proven or accepted formally as a postulate. Since it is a reasonable statement, no elaboration seems needed at this point.

(c) Draw diameter $\overline{XC}$. Then use case (a) on $\angle AXC$ *and* $\angle BXC$ and then subtract.

Perhaps the most notable case of the above theorem occurs when the angle is inscribed in a semicircle, in which case it is, of necessity, a right angle.

EXERCISE SET 11.3

1. If a quadrilateral is inscribed in a circle, what is true about the opposite angles of the quadrilateral?

2. If a parallelogram is inscribed in a circle, what is true about this figure?

3. Prove: In a circle, two congruent chords have congruent arcs.

4. State and prove the converse of the statement in Exercise 3.

5. In the figure below, $\overline{AB} \parallel \overline{CD}$. Prove that minor arcs $\overset{\frown}{AC}$ and $\overset{\frown}{BD}$ have the same measure.

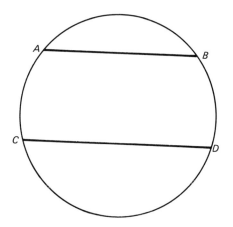

6. As found in Exercise 7 of the previous Exercise Set, regular polygons can always be inscribed in a circle. Envision a regular hexagon inscribed in a circle. If radii are drawn to the six vertices, six triangles are formed. What kind of triangles are they? What is the measure of each of the interior angles of the hexagon? Answer the same questions for a square.

11.4 MEASUREMENT

As was the case with the other geometric figures studied, there is interest in finding measures associated with circles and spheres. For the circle, both the distance around (or the length) as well as the area of the circular region associated with it are worth knowing for their physical applications. For the sphere, the surface area and volume are the measures of interest.

Beginning with the circle, let us consider its length, which is called the **circumference**, usually symbolized by C. A circle is a curve of finite length, and the measure of its length can be found by a general method whose ultimate solution rests upon the limit concept of calculus. Although these notions are beyond the scope of this text, its basic ideas can be made intuitively clear.

As an example of a limit, consider the sequence of rational numbers,

$$\frac{1}{2}, \frac{1}{4}, \frac{1}{8}, \frac{1}{16}, \frac{1}{32}, \frac{1}{64}, \frac{1}{128}, \frac{1}{256}, \cdots$$

If we were to add these one at a time, the successive sums would be

$$\frac{1}{2}, \frac{3}{4}, \frac{7}{8}, \frac{15}{16}, \frac{31}{32}, \frac{63}{64}, \frac{127}{128}, \frac{255}{256}, \cdots$$

Each term of this sequence of sums gets larger every time a new number is added, but yet each term always remains less than 1. We say that the number "1" is the limit of this sequence of sums as the number of terms of the original sequence being added becomes large without bound.

In a similar way, consider an arc of a curve as pictured in Fig. 11-11. By choosing points A, B, C, D, E, F, and G, and then drawing segments $\overline{AB}$, $\overline{BC}$, $\overline{CD}$, $\overline{DE}$, $\overline{EF}$, and $\overline{FG}$, we get a succession of segments whose lengths can be determined by known methods. The sum of the measures of these segments is a reasonable and close approximation to the length of the arc. If additional points were added, as shown in Fig. 11-12, at X and Y, a new sum of measures of segments is derived that is greater than or equal to the previous sum, since

$$m(\overline{CX}) + m(\overline{XD}) \geq m(\overline{CD}) \quad \text{and} \quad m(\overline{EY}) + m(\overline{YF}) \geq m(\overline{EF})$$

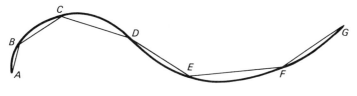

Figure 11-11

By increasing the number of points and consequently making each individual segment smaller, we increase the sum of the measures of the segments. We say that the limit of this sum is the length of the arc. The methods of finding this limit is the content of the calculus.

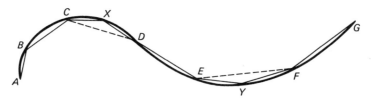

Figure 11-12

For the circle, the best method of using this limit concept is to imagine a regular polygon of n sides inscribed in a circle. The perimeter of this polygon is the succession of segments, the sum of whose measures is the approximation to the circumference of the circle. Now double the number of sides by finding the midpoint of each arc of the circle and join the adjacent vertices of the polygon to these points. Figure 11-13 illustrates this concept, starting with a square and then proceeding to an octagon and then to a polygon with 16 sides. If we repeat this process infinitely many times, the perimeter gets closer and closer to the actual value of the circumference. Symbolically, we write:

$$p \to C \quad \text{as} \quad n \to \infty$$

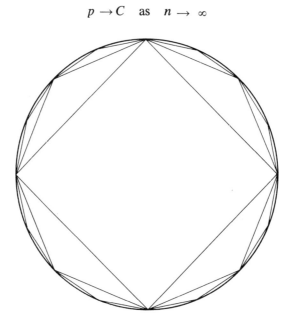

Figure 11-13

Without any derivation, the formula for the circumference of a circle is given:

$$C = \pi d$$

where π is an irrational number approximated by 3.14159265358979. This approximation is a close one; some of the more common approximations are 3.14, 3.1416, and $\frac{22}{7}$.

When the formula is written as $C/d = \pi$, the fact that the ratio of the measure of the circumference to the measure of the diameter is a constant is brought out. This is a very important fact. The ancient Greeks were aware of this fact, although they did not calculate the value of π to the great accuracy that is known today. Physical measurements with circular-shaped objects can lead school children to make the conjecture that the ratio is a constant. Careful measurement and averaging can often lead to a sur-prisingly close approximation.

Since the diameter of a circle is equal to two radii, the above formula is often written as:

$$C = 2\pi r$$

With the above facts in mind, we find it a relatively easy matter to derive the formula for the area of a circular region. Using the idea of the inscribed regular polygon with an ever-increasing number of sides, we again make use of the limit concept. Recall that the formula for the area of a regular polygonal region is:

$$A = \tfrac{1}{2}a \cdot p$$

where a is the measure of the apothem and p is the perimeter. We have seen that $p \rightarrow C = 2\pi r$, as n gets large. What happens to a as n gets large? It gets larger and has r as its limit. Putting these together, we see that

$$A = \tfrac{1}{2}a \cdot p \rightarrow \tfrac{1}{2}(r) \cdot (2\pi r) = \pi r^2 \quad \text{as } n \text{ gets large}$$

This result, then, yields the correct formula for the area of a circular region:

$$A = \pi r^2$$

Due to their complicated nature, the derivations of the formulas for the surface area and volume of a sphere will be given here without any appeal to either intuition or proof.

The surface area of a sphere, with radius r, is given by the formula:

$$A = 4\pi r^2$$

The volume of a sphere, with radius r, is given by the formula:

$$V = \tfrac{4}{3}\pi r^3$$

EXERCISE SET 11.4

1. Complete the following table using $\frac{22}{7}$ as an approximation for π.

   (a) For a circle:

r	d	C	A
14 in.	—	—	—
—	2.8 cm	—	—
—	—	11 ft	—
—	—	—	55.44 sq in.

   (b) For a sphere:

r	d	A	V
4 cm	—	—	—
—	6 cm	—	—
—	—	$\frac{7}{22}$ sq in.	—

2. What is the distance around the equator of the earth if the earth is considered to be a spherical solid with a diameter of 8,000 mi?

3. Suppose the earth was bounded by a snug fitting metal band around its equator. If the band were lengthened by one foot, would the slack in the snugness (if taken up evenly around the distance of the equator) be enough to enable a piece of paper to slide underneath it? For a mouse to crawl under it? For a human being to wiggle under it?

4. (a) In the figure below, arc $\overset{\frown}{AB}$ is a semicircle. P is the midpoint of $\overline{AB}$, and the smaller arcs are semicircles with $\overline{AP}$ and $\overline{PB}$ as their

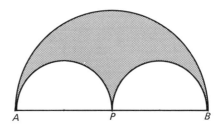

diameters. What is the area of the shaded region in terms of a larger semicircular region? Before "figuring it out," make an "educated guess." (b) Is the point P of part (a) a special point? That is, would the result of part (a) be the same if P were not the midpoint of $\overline{AB}$? The figure below shows another position for P. Q is the midpoint of $\overline{AB}$. Again, make an "educated guess" before working out the answer.

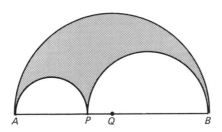

11.5 CYLINDERS AND CONES

The familiar "tin can" in which beverages and canned foods are packaged is a physical model of a geometric object known as a cylinder. As such, it is a special case of a simple, closed surface. However, the "tin can"-shaped cylinder is but a special case of the cylinder, which shall be defined below.

Consider a curve lying in a plane α and a line, $\overleftrightarrow{AB}$, that intersects α in exactly one point. The set of all lines parallel to $\overleftrightarrow{AB}$ and containing a point of the curve is known as a **cylindrical surface**. If the curve is a simple closed curve, then the surface is called a **closed cylindrical surface**. The curve is called the directrix of the surface, and any of the lines mentioned in the description above is called a **generator** of the surface. These latter terms suggest a way of thinking of a cylindrical surface as a set of points "generated" by a line "moving" along the curve sweeping out the surface This concept is illustrated in Fig. 11-14.

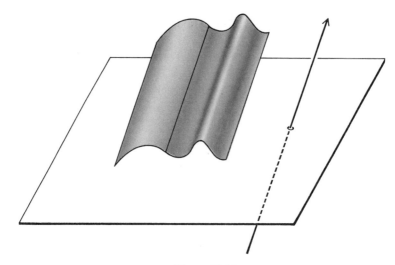

Figure 11-14

DEFINITION 11.15

A **cylinder** is a set of points that is the union of points on a closed cylindrical surface lying between two parallel planes that intersect the surface in two congruent simple closed curves, with the points of the two planes that lie in the region determined by the two simple closed curves.

The definition sounds complicated, but the idea is simple. A cylinder (see Fig. 11-15) has two congruent bases lying in two parallel planes and a lateral surface. It is akin to a prism; in fact, if the simple closed curve is a polygon, then the cylinder is a prism. Thus, we see that a prism is a special case of a cylinder.

Figure 11-15

If the base is a circular region, then the cylinder is a **circular** cylinder. A cylinder is a **right** cylinder in exactly the same way that a prism is a right prism. The familiar "tin can," then, is a **right circular cylinder.** The segment connecting the centers of the two circular bases is called an **axis** of the cylinder, and, if the cylinder is a right circular cylinder, the axis is perpendicular to the planes of the bases.

The volume of a cylindrical solid can be found by using exactly the same formula as applicable to the prism, $V = Bh$, where B is the area of the base and h is the measure of the altitude. Altitude has the meaning here as it did for the prism. If the cylinder is a circular cylinder, the formula may take on a special form, incorporating the formula for the area of a circle. The formula is:

$$V = \pi r^2 h$$

where r is the radius of the circle that determines the base.

The **cone** is the generalization of the pyramid in exactly the same way that the cylinder is the generalization of the prism. Again, we start with a curve lying in a plane α. Now consider a point P not lying in α. The set of all lines lying on P and containing a point of the curve is called a **conical surface.** The curve is called the **directrix** and the fixed point, the **vertex** of the cone. As the description suggests and as Fig. 11-16 shows, there are two parts to a conical surface that are symmetrical to each other. Each is called a **nappe** of the surface. If the curve is a simple closed curve, the surface is a **closed conical surface.**

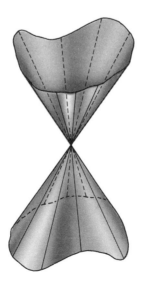

Figure 11-16

Since we know that a pyramid is a special case of a simple closed curve that is called a cone, it should be easy enough to formulate a definition for a cone. This formulation is left as an exercise.

The simple closed region lying in the plane of the definition of a cone is the **base** of the cone; if it is a circular region, then the cone is a **circular cone**. If the segment connecting the vertex of a cone to the center of a circular base is perpendicular to the plane of the base, then the cone is a **right circular cone**.

By analogous reasoning, the formula for the volume of a conical solid is the same as for the pyramid, $V = \frac{1}{3}Bh$, where B is the area of the base and h is the measure of the altitude. (What is the altitude of a cone?) If the base is a circular region, then the special form of the formula is:

$$V = \tfrac{1}{3}\pi r^2 h$$

where r is the radius of the circle that determines the base.

EXERCISE SET 11.5

1. Formulate a precise definition for a cone. (*Hint*: Use only one nappe of the conical surface.)

2. The lateral area of a circular cylinder is the area on the cylindrical surface lying between the two bases. If the radius of the base is r and the height of the cylinder is h, what is the measure of the lateral area?

3. To find the lateral area of a cone, imagine the cone as being formed by a piece of paper, which is then cut along a segment $\overline{PQ}$ (called the **slant height**), as shown in the figure, and then laid out flat on a table to form the shape shown below as the shaded area. Let r be the radius

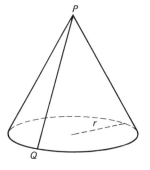
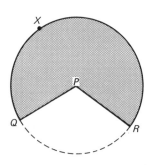

of the base of the cone. Can you determine a formula for the lateral area of the cone? (*Hint:* Express the length of $\overparen{QXR}$ in terms of r. Can the length of this arc be related to the total area of the circle with center at P with radius $\overline{PQ}$?)

4. Complete the following tables. (Do not use an approximation for π, but rather leave the symbol "π" in your answers.)

   (a) For a right circular cylinder:

r	h	Lateral area	Total surface area	V
3	2	—	—	—
2	2	—	—	—
—	4	24 π sq units	—	—

   (b) For a right circular cone:

r	h	Lateral area	Total surface area	V
3	4	—	—	—
2	2	—	—	—
—	2	—	—	6π cu units

5. Imagine a ball of ice cream in the shape of a sphere being placed on top of a right circular cone (with base removed), so that the radii of the sphere and the base of the cone were equal. If the ice cream were allowed to melt completely, it would turn into a liquid that would then fill the interior of the cone. What would the height of the cone have to be in order to keep the liquid from overflowing?

6. A **frustum** of a right circular cone is a solid formed by "chopping off" the top of a conical solid by a plane parallel to the base of the cone.

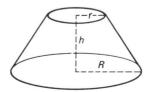

The upper base of the frustum is a circular region with radius smaller then the radius of the base. Find the volume of a frustum of a cone if the radius of the lower base is R, the radius of the upper base is r, and h is the measure of the altitude between the two bases.

7. Describe all possible intersections between a plane and a right circular cylinder.

8. Describe all possible intersections between a plane and a right circular cone.

SUGGESTED READINGS

[11] Ch. 9

[22] Ch. 11, 12

[24] Ch. 11

[26] Ch. 5

12

COMPASS AND STRAIGHTEDGE CONSTRUCTIONS

It is almost a tautology to say that drawings of geometric figures help a person gain a better understanding of the properties of those figures. Certain construction problems often help students to gain an appreciation and insight into the subject, many times because they themselves discover some property. The classical instruments of geometric construction problems are the unruled **straightedge,** which permits the drawing of a line—actually only a segment since a line is infinite in length—and the **compass,** or dividers, which permits the drawing of a circle or arc thereof. Originally, the compass was a "collapsible" instrument, which meant that when it was lifted from the paper, the two points of the compass would fall towards each other, and the user was unable to preserve the distance between the two points. Nowadays, the compass is of the rigid variety, which enables us to pick up the compass and "transfer" the distance between the endpoints of some segment to another location on a piece of drawing paper.

The geometry of this text has been, for the most part, of the metric variety, since we have used coordinate lines in order to give segments a measure and also coordinated half-planes in order to give measures to angles.

Thus, the ordinary ruler and protractor would be permissible instruments. The use of such instruments would not make the construction "problems" problems in the true sense of the word. There would be no difficulty in drawing figures. However, since it is of some historic interest to present the construction using only the straightedge and compass, the present chapter will give a short description of the basic constructions needed in a beginning course in geometry. Other construction problems involve only a combination of the basic techniques.

12.1   SOME BASIC IDEAS

The basic constructions will involve the drawing of segments and circles. A segment is completely determined by its endpoints, and a circle is determined by its center and the length of its radius, which is a segment. Thus, the entire question rests upon the location of points. Some of the familiar geometric figures are sets of points with particular properties. For example, a circle is a set of points, all of which are the same distance from a point called the center of the circle; we have seen that an angle bisector is a ray, all of whose points (except the endpoint) are equidistant from the sides of the angle. The intersection of such sets of points would be a set of points that possess the properties of both original sets of points.

Since the instruments allowable can only draw segments and circles, we should investigate the possible intersections of such figures. Two segments, if they intersect at all, will intersect in only one point, if they are distinct segments. A segment and a circle will not intersect at all, or will intersect in one point (if the segment is tangent to the circle), or in two distinct points. (Since the segment is the drawing "idealization" of a line, we may consider it as extending as far as needed. Thus, a segment having an endpoint in the interior of the circle, and, thus, having only one point in common with the circle, is not considered a possibility.)

Two circles will intersect, depending upon the length of the segment joining their centers in comparison to the sum of the measures of the radii. Depending on whether the measure of the segment is greater than, equal to, or less than the sum of the measures, the circles will interest in no, one, or two points.

12.2 THE COPYING CONSTRUCTIONS

The first construction is to copy a given segment. To put it more precisely, the problem is, given a segment $\overline{AB}$, to find a point X on a given ray $\overrightarrow{PQ}$ such that $\overline{PX} \cong \overline{AB}$. (Recall that we postulated in Chapter 4 that it was always possible to find such a point.)

To perform this construction, use the compass so that the sharp point is placed on point A of $\overline{AB}$ and then open the compass so that the pencil point is placed on B. Then lift the compass to place the sharp point on P. With the compass so placed, draw an arc of a circle with radius $\overline{AB}$ so that the arc intersects the given ray $\overrightarrow{PQ}$. The intersection of the ray with the arc is the desired point X.

Copying a Segment

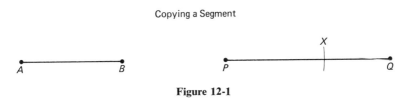

Figure 12-1

The other copying construction is the copying of an angle, so that its vertex lies at a given point on a given line and the other side lies in a particular half-plane of that line. To put it more precisely again, the problem is, given a fixed angle, $\angle ABC$, a fixed line $\overleftrightarrow{PQ}$ and a point X not on $\overleftrightarrow{PQ}$, to find a point M on the X-side of $\overleftrightarrow{PQ}$ such that $\angle MPQ \cong \angle ABC$.

Begin by placing the sharp point of the compass on the vertex of $\angle ABC$. Opening the compass to any length, say $\overline{BE}$, draw an arc of a circle with radius $\overline{BE}$ so that the arc intersects rays $\overrightarrow{BA}$ and $\overrightarrow{BC}$, as shown in Fig. 12-2.

Copying an Angle

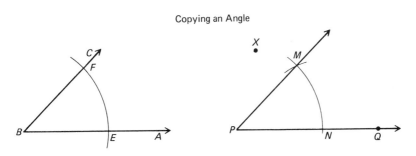

Figure 12-2

265

Then, with the same length, place the sharp end at P and draw an arc of a circle with radius $\overline{BE}$ so that the arc intersects $\overset{\leftrightarrow}{PQ}$ at some point N and so that the arc lies in the X-side of $\overset{\leftrightarrow}{PQ}$. Now, with the compass, place the sharp point at E and open up the compass so that the other point is placed on F. This "gives" you segment $\overline{EF}$. Now lift the compass and place the sharp end at N and draw an arc of a circle with radius $\overline{EF}$ so that the arc intersects the previously drawn arc with center P and radius $\overline{BE}$. These arcs intersect at the desired point M.

Using the compass to transfer segments is considered to be a legitimate way of demonstrating that two segments are congruent. In fact, for younger children, such a technique might well be considered a valid way of showing congruence. (For their teacher, it may very well be a satisfactory way of defining congruence.) With this in mind, it is easy to prove that the construction "works." Although segments $\overline{EF}$ and $\overline{NM}$ are not shown in Fig. 12-2, consider the triangles, $\triangle BEF$ and $\triangle PNM$.

$$\overline{BE} \cong \overline{PN}, \ \overline{BF} \cong \overline{PM}, \text{ and } \overline{EF} \cong \overline{NM}$$

all by construction. Thus, the two isosceles triangles are congruent by the SSS Postulate, and the angles are congruent as a result.

With the above two constructions, the copying of triangles is an easy matter. If the SAS Postulate is the desired rationale for copying, we would start by copying a segment, then copying the angle at the desired end of that segment, and complete the construction by copying the second segment on the second side of the just-copied angle with its vertex as the given point. If the ASA Postulate is the rationale, copy the side first and then the two angles on either end of that segment.

To copy a triangle using the SSS rationale, consider the triangle ABC of Fig. 12-3. On a given line, copy $\overline{AB}$, calling it $\overline{DE}$. Using the compass, transfer $\overline{AC}$ by placing the sharp point at D and drawing an arc of a circle with radius $\overline{AC}$; then do the same with $\overline{BC}$ by placing the sharp point at E and drawing an arc of a circle with radius $\overline{BC}$. The intersection of these two arcs (there will be two such points—one on each side of $\overset{\leftrightarrow}{DE}$) is a point F such that $\triangle ABC \cong \triangle DEF$.

Copying a Triangle by "SSS"

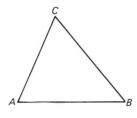
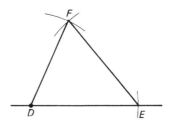

Figure 12-3

12.3   THE BISECTING CONSTRUCTIONS

The first construction of this type is the construction of a ray in the interior of a given angle, so that the ray is the bisector of the given angle. To be more precise, the problem is, given an arbitrary angle, $\angle ABC$, to find a point P in the interior of $\angle ABC$ such that $\angle ABP \cong \angle CBP$.

Placing the sharp point at the vertex of the angle, select a radius of any convenient length, say $\overline{BE}$, and draw an arc of a circle with center at B and radius $\overline{BE}$ so that the arc intersects the two sides of the angle at points E and F. Using E and F as centers, draw arcs of two congruent circles of any convenient radius so that the arcs intersect in the interior of the angle. The intersection of the arcs is the desired point P.

Bisecting an Angle

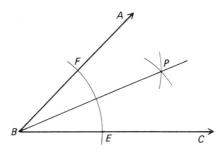

Figure 12-4

It is easily shown that P is the desired point. $\overline{BE} \cong \overline{BF}$ and $\overline{EP} \cong \overline{FP}$, both by construction. Certainly, $\overline{BP} \cong \overline{BP}$. Thus, $\triangle BEP \cong \triangle BFP$ by SSS. The desired congruence of the angles follows directly.

The other bisection problem is to find the midpoint of a segment. To put it more precisely, the problem is, given a segment $\overline{AB}$, to find a point P of $\overline{AB}$ such that $\overline{AP} \cong \overline{PB}$.

Place the sharp point of the compass on B and draw a large arc of radius larger than half of segment $\overline{AB}$. (This can always be done by choosing $\overline{AB}$ as the length of the radius.) Then, with the sharp point at A, draw a large arc of a circle with radius equal to the radius of the first drawn arc so that the arcs intersect in two distinct points, one on one side of $\overleftrightarrow{AB}$ and the other on the opposite side of $\overleftrightarrow{AB}$. Call these two points M and N. The intersection of $\overline{MN}$ and $\overline{AB}$ is the desired point P. With a little practice and some visual estimation, the arcs do not have to be drawn as large as is indicated in Fig. 12-5; only the portion of the arcs near the points M and N need be drawn. The important thing to remember is that all of these arcs have the same radius.

Bisecting a Segment

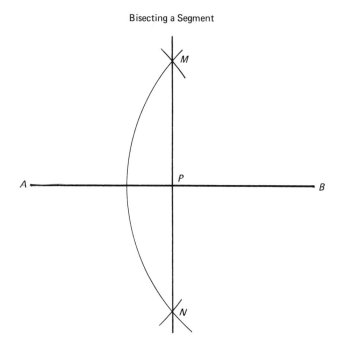

Figure 12-5

The proof that P is the midpoint of $\overline{AB}$ is left to you. As a matter of fact, this construction gives the added "bonus" of the line $\overleftrightarrow{MN}$ being perpendicular to $\overline{AB}$ as well as intersecting $\overline{AB}$ at its midpoint.

12.4 THE LINE CONSTRUCTIONS

Of interest here is the construction of lines perpendicular and parallel to given lines. Consider the case of the perpendicular first. The problem is, given a line $\overleftrightarrow{AB}$ and a point P either on or off $\overleftrightarrow{AB}$, to find a point Q such that $\overleftrightarrow{PQ} \perp \overleftrightarrow{AB}$.

Using P as the center, draw two arcs of the same circle with suitable radius (if P is not a point of $\overleftrightarrow{AB}$, the radius must be longer than the "distance" from P to $\overleftrightarrow{AB}$) so that the arcs intersect $\overleftrightarrow{AB}$ in two distinct points, say M and N. With M and N as centers, draw arcs of two congruent circles of any convenient radius so that the arcs meet on the side of $\overleftrightarrow{AB}$ opposite from the P-side. (If P lies on $\overleftrightarrow{AB}$, then either side of $\overleftrightarrow{AB}$ will do.) The intersection of the two arcs is the desired point Q.

A Perpendicular to a Line through a Given Point

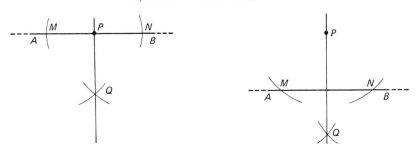

Figure 12-6

In part (a) of Fig. 12-6, $\overline{MP} \cong \overline{NP}$ and $\overline{MQ} \cong \overline{NQ}$ both by construction. $\overline{PQ} \cong \overline{PQ}$, and so $\triangle MPQ \cong \triangle NPQ$ by SSS. This leads to the two angles at P being congruent, and since they are a linear pair, they are right angles and the lines are perpendicular. In part (b), $\overline{MP} \cong \overline{NP}$ and $\overline{MQ} \cong \overline{NQ}$, both by construction. At this point, the proof becomes identical with the proof of the segment bisector of the previous section, if one considers $\overline{MN}$ as the segment. In that proof, the by-product was that $\overleftrightarrow{PQ} \perp \overleftrightarrow{MN} = \overleftrightarrow{AB}$.

269

For a parallel to a given line, the problem is, given a line $\overleftrightarrow{AB}$ and a point P not on $\overleftrightarrow{AB}$, to find a point Q such that $\overleftrightarrow{PQ} \parallel \overleftrightarrow{AB}$.

Through P draw any line that intersects $\overleftrightarrow{AB}$ in some point X where $X \neq A$. Thus, $\angle PXA$ is formed. Now, simply copy $\angle PXA$, using P as the vertex, $\overrightarrow{PX}$ as the given ray, so that the other side of the angle lies in the half-plane opposite to the A-side of $\overleftrightarrow{PX}$.

In this construction, the lines are parallel because congruent alternate interior angles have been constructed.

A Parallel to a Given Line through a Given Point

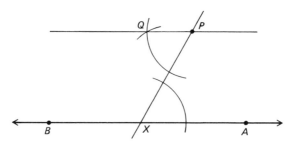

Figure 12-7

EXERCISE SET 12.1

1. Construct an isosceles right triangle.

2. Construct a 30-60-90 triangle.

3. For an arbitrarily drawn triangle ABC, construct the inscribed circle.

4. For an arbitrarily drawn triangle ABC, construct the circumscribed circle.

5. For an arbitrarily drawn triangle ABC, construct the three altitudes.

6. For an arbitrarily drawn circle, construct its inscribed regular hexagon.

7. For an arbitrarily drawn circle, construct its circumscribed regular hexagon.

8. For an arbitrarily drawn circle, construct its inscribed square.

9. Draw an arbitrary segment and let it have unit length. With this segment of length 1 as your beginning point, construct a segment of length $\sqrt{2}$, and then a segment of length $\sqrt{3}$. (*Hint:* Consider the Pythagorean theorem.)

10. Given two segments with lengths a and b, where $a > b$. Construct an isosceles trapezoid with a and b as lower and upper bases, respectively, and having lower base angles with a measure of 30.

11. Given a segment $\overline{AB}$. Construct the set of all points P such that $m(\angle APB) = 90$.

12. Given a circle with center at P and an external point O. Construct one of the tangents from O to circle P.

SUGGESTED READINGS

[12]

[17]

[28]

13

COORDINATE GEOMETRY IN THE PLANE

In an earlier chapter, the idea of coordinates of points on a line was introduced. In the present chapter, this notion is extended to the two-dimensional realm of the plane. In fact, we could extend it even further to the three dimensions of space; however, this will not be done in this text. By giving numerical coordinates to points of the plane, we form an extremely close bond between the central ideas of geometry and the fundamental techniques of algebra. In this study of coordinate geometry, the algebraic equation plays a vital role.

13.1 COORDINATES OF A POINT

Because the points of the plane need not all lie on a single number line, something more than a single number is needed to identify and locate these points. Several methods have been devised to impose a coordinate system on the plane. The one that will be studied was developed by René Descartes, a French philosopher and mathematician of the seventeenth century who, some say, got his idea for his coordinate system from observing the windmills of Holland while he was a visitor in that country.

The heart of the system is two coordinate lines, placed at right angles to each other with the assumption that the positive direction on each line is known. Although they may lie in any position, the two lines are normally depicted as being horizontal and vertical, as shown in Fig. 13-1. With the lines lying in these positions, the positive directions for the lines are normally taken as to the right for the horizontal line and upwards for the vertical line.

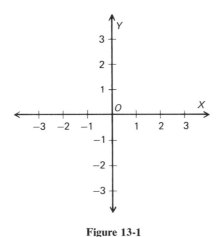

Figure 13-1

As a further convention, the horizontal line is generally called the **x-axis** and the vertical line is called the **y-axis**. The point of intersection of the two coordinate axes is called the **origin** and is labeled by the letter O. As such, the origin is the origin of each of the two lines and is the zero point of each. The coordination of each axis is made with respect to the same unit pair; thus, each axis is said to have the same scale, which should be indicated whenever a drawing is made. This same coordination is not absolutely essential, but without it, some strange figures could occur, such as the odd looking "square" of Fig. 13-2.

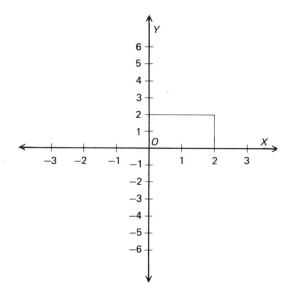

Figure 13-2

Because the plane is a two-dimensional object, the points of the plane are characterized by a pair of numbers; this characterization is done in the following manner.   Consider a point P of the plane under consideration. From P, draw perpendiculars to the x-axis and the y-axis, meeting those lines at points X and Y, respectively.   The coordinate of X on the x-axis is called the **abscissa** (or x-coordinate) of P, and the coordinate of Y on the

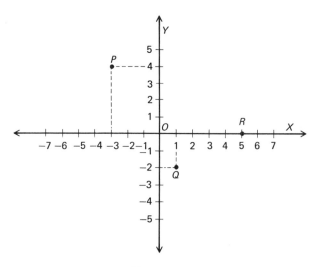

Figure 13-3

y-axis is called the **ordinate** (or *y*-coordinate) of *P*; together, the pair of numbers are called the **coordinates** of *P*.   The coordinates of any point in the plane may be found in this manner.   For convenience, the coordinates of a point are listed together, surrounded by parentheses and separated by a comma.   So that there be no confusion as to the identity of the coordinates, ordinary practice stipulates that the *x*-coordinate be listed first and the *y*-coordinate second.   Thus, in Fig. 13-3, point *P* has the coordinates $(-3, 4)$, while *Q* has the coordinates $(1, -2)$, and *R* has the coordinates $(5, 0)$. These facts will be written as $P:(-3, 4)$, $Q:(1, -2)$; and $R:(5, 0)$.

Conversely, if one is given the ordered pair $(2, 3)$, for example, he can locate the point corresponding to that pair of numbers by erecting perpendiculars to the *x*- and *y*-axes at the points whose coordinates are 2 and 3, respectively, on those lines.   The intersection of those perpendiculars is the desired point.   Figure 13-4 illustrates this concept.

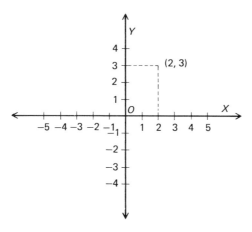

Figure 13-4

The preceding paragraphs lead to the following postulate.

POSTULATE 13.1

Given a plane α, there exists a one-to-one correspondence between the points of that plane and the set of all ordered pairs of real numbers.

This correspondence is called a **coordinate sysem.**   Because of this one-to-one correspondence, one often sees or hears the use of such imprecise, yet convenient statements as "the point $(3, 4)$" or "the point (x_0, y_0)." Such a practice will be adopted in this text.

The intersections of the half-planes determined by the coordinate axes are known as **quadrants.** They are numbered by Roman numerals, starting with the quadrant in which both coordinates are positive and then working in a counterclockwise direction.   Figure 13-5 illustrates the naming of the four quadrants.

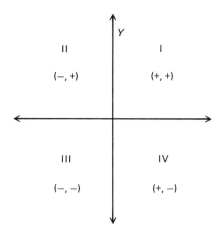

Figure 13-5

EXERCISE SET 13.1

1. Give the coordinates of the points pictured below.

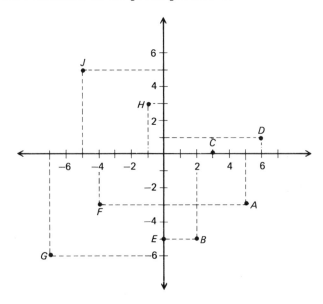

2. What are the coordinates of the origin?

3. What is the ordinate of all points lying on the x-axis? What is the abscissa of every point on the y-axis?

4. What are the coordinates of the point that is the intersection of the y-axis and the perpendicular from $P:(-5, 9)$ to the y-axis?

5. The lines through $Q:(1, 7)$ perpendicular to the coordinate axes form a rectangle. What is the perimeter and area of this rectangle? Find the same for $S:(-2, -3)$; for $M:(5, -4)$.

6. If $A:(-3, 2)$, $B:(-3, -3)$, and $C:(5, 2)$ are three of the four vertices of a rectangle, what are the coordinates of D, the fourth vertex? What is the perimeter and area of this rectangle?

7. If $M:(2, -7)$, $N:(5, -7)$, and $P:(9, -4)$ are three of the four vertices of parallelogram $MNPQ$, what are the coordinates of the fourth vertex? What is the perimeter and area of this parallelogram?

8. Describe another method of establishing a coordinate system for a plane. (*Hint:* Consider the "quasi" coordination of a half-plane described in the section on finding angle measure.)

9. Describe how one might go about setting up a coordinate system for space.

10. With reference to Postulate 13.1, could there be more than one one-to-one correspondence between points of a plane and the set of all ordered pairs of real numbers? Explain your answer.

13.2   SLOPE OF A LINE

Since the straight line, and subsets thereof, are the building blocks of geometric figures in the plane, it is necessary to have some device for characterizing and identifying them. Refer to Fig. 13-6. Consider a fixed point A on the x-axis with the coordinates $(a, 0)$, where a stands for any real number. On this point of the plane, there lie infinitely many lines. However, each individual line may be identified by means of the angle determined by it and the x-axis. For purposes of explanation, consider one such line that lies on the point $P_0:(x_0, y_0)$. Call the foot of the perpendicular from P_0 to the x-axis, X_0. The angle, $\angle P_0 A X_0$, has a unique measure; therefore, the line $\overline{AP_0}$ is completely characterized and described, if we know the measure of that angle.

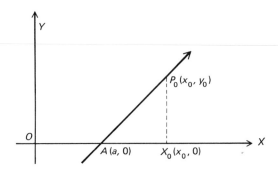

Figure 13-6

Recalling the trigonometric ratios and, in particular, the tangent of an angle, we note that these ratios are unique for each angle of distinct measure. In essence, this fact enables us to describe the measure of the angle in terms of one of these ratios; as will be presently shown, the ratio in question may be described by referring to the coordinates of two points of the line.

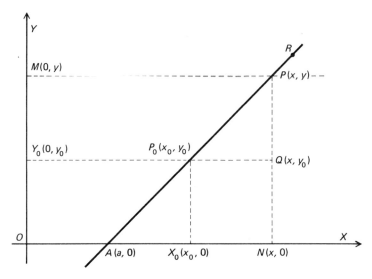

Figure 13-7

The trigonometric ratio of interest is the tangent. If we refer to Fig. 13-7, the tangent of $\angle A$, when considering $\triangle AP_0X_0$, is $\dfrac{P_0X_0}{X_0A}$. The length of $\overline{P_0X_0}$ is the same as that of $\overline{Y_0O}$. Since $\overline{Y_0O}$ is a segment of a number line—the y-axis—its length is found to be $|(y_0 - 0)|$. The length of $\overline{X_0A}$ is $|(x_0 - a)|$. Therefore, the tangent may be written as $\dfrac{|y_0 - 0|}{|x_0 - a|}$. However, if one considers any other point, say $P:(x, y)$, and draws the appropriate segments, a set of similar triangles is formed. Thus, $\triangle AP_0X_0 \sim \triangle APN \sim \triangle P_0PQ$. The resulting equal ratios from these similarities are:

$$\frac{\overline{P_0X_0}}{\overline{X_0A}} = \frac{\overline{PN}}{\overline{NA}} = \frac{\overline{PQ}}{\overline{QP_0}}$$

When these equalities are expressed in terms of coordinates, we have:

$$\frac{|y_0 - 0|}{|x_0 - a|} = \frac{|y - 0|}{|x - a|} = \frac{|y - y_0|}{|x - x_0|}$$

Because of these equalities, any one of the ratios of the difference between coordinates could be used to describe the line $\overleftrightarrow{AP}_0$. (Note that $\overleftrightarrow{AP}_0 = \overleftrightarrow{AP} = \overleftrightarrow{PP}_0$.) Because these ratios do enable us to characterize a line by means of coordinates of points on that line rather than by referring to an angle, the term tangent is not used in this connection. Instead, the term *slope of a line* is introduced.

DEFINITION 13.1

The **slope of a line** $\overline{P_1P_2}$, symbolized by $m_{P_1P_2}$, is given by the ratio $m = \dfrac{y_2 - y_1}{x_2 - x_1}$, where $P_1:(x_1, y_1)$ and $P_2:(x_2, y_2)$.

The quantity $(y_2 - y_1)$ is called the *change in y* from P_1 to P_2 and is often symbolized by $\triangle y$. (Read: delta y.) Similarly, the quantity $(x_2 - x_1)$ is called the *change in x* from P_1 to P_2 and is symbolized by $\triangle x$. This symbolism is illustrated in Fig. 13-8. Thus, if one "moves" along a line from P_1 to P_2 so that the horizontal component of the move is in the positive direction, $\triangle x$ is a positive quantity; conversely, if the horizontal component is in a negative direction, $\triangle x$ is a negative quantity. Similar statements hold for $\triangle y$.

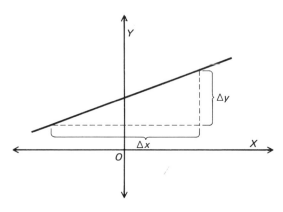

Figure 13-8

Therefore, the slope is often written as $m = \triangle y / \triangle x$. To help remember this ratio, the slope is verbally described as "rise over run." Since

$$\frac{y_2 - y_1}{x_2 - x_1} = \frac{y_1 - y_2}{x_1 - x_2}$$

it makes no difference which of two given points is called the "first" point and which is the "second." The thing that does matter is that the two coordinates of one point be listed first in the ratio of the differences of coordinates. In other words, $\dfrac{y_2 - y_1}{x_1 - x_2}$ would not give a correct value for the slope of a given line.

EXAMPLES

(a) What is the slope of the line passing through $A:(-2, 3)$ and $B:(4, 1)$?

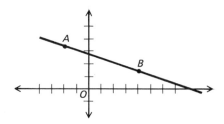

$$m_{AB} = \frac{(3-1)}{(-2-4)} = \frac{(1-3)}{(4-(-2))}$$

$$= \frac{2}{-6} = \frac{-2}{6} = -\frac{1}{3}$$

(b) What is the slope of $\overleftrightarrow{MN}$, where $M: (-1, -5)$ and $N:(3, 0)$?

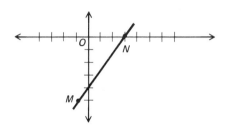

$$m_{MN} = \frac{(-5-0)}{(-1-3)} = \frac{(0-(-5))}{(3-(-1))}$$

$$= \frac{-5}{-4} = \frac{5}{4}$$

(c) What is the slope of $\overleftrightarrow{RT}$, where $R:(3, 2)$ and $T:(3, -1)$?

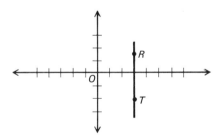

$$m_{RT} = \frac{(2-(-1))}{(3-3)} = \frac{3}{0}$$

which is undefined. $\overleftrightarrow{RT}$ is parallel to the y-axis and is called a vertical line. *All vertical lines have no slope.*

(d) What is the slope of $\overleftrightarrow{JK}$, where $J:(-6, 4)$ and $K:(1, 4)$?

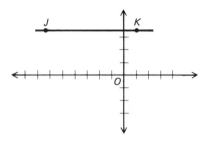

$$m_{JK} = \frac{(4 - 4)}{(-6 - 1)} = \frac{0}{-7} = 0$$

$\overleftrightarrow{JK}$ is parallel to the x-axis and is called a horizontal line. *All horizontal lines have a slope equal to zero.*

Any segment or ray is said to have a slope equal to the slope of the line of which it is a subset.

In view of the previous paragraphs, the following theorem may be stated without any further proof.

THEOREM 13.2

Two lines are parallel if and only if they have equal slopes.

Example (b) above dealt with a line whose slope was a positive number. All lines that have a positive slope are positioned in relatively the same manner; that is, if viewed from left to right, the line rises upward. Example (a) dealt with a line with a negative slope, and as the accompanying figure shows, the line slopes downward when viewed from left to right.

It is also of importance to be able to describe perpendicular lines in terms of their slopes. By definition, horizontal and vertical lines are perpendicular; thus, to avoid the numerical complications that go with the slopes of those lines, the discussion will consider only nonvertical and non-horizontal lines. Intuitively, it is clear that the slopes of two perpendicular lines must have opposite signs—one must be positive and the other, negative. This condition and one other that is needed for perpendicularity are stated in the following theorem.

THEOREM 13.3

Two nonvertical lines l_1 and l_2, with slopes m_1 and m_2, respectively, are perpendicular if and only if $m_1 = -1/m_2$.

The proof of this theorem must be postponed until the next section after one further result is given.

Returning to the original discussion that was given to motivate the definition of slope, we shall see that there is another rich "bonus" awaiting us. The discussion began with point A on the x-axis; however, the ensuing paragraphs indicated that point A is not really special—any point on that line serves the same purpose, as indicated by the fact that the angles, $\angle PP_0Q$ and $\angle RPT$, are congruent to $\angle PAN$. In other words, a line may be completely determined by knowing its slope and any point of that line. This fact can be expressed algebraically as follows.

Suppose there is a fixed line, with slope m, lying on a specific point $P:(x_0, y_0)$. Then, for any other point $P:(x, y)$ on that line, we have:

$$m = \frac{y - y_0}{x - x_0} \tag{13-1}$$

which is equivalent to

$$(y - y_0) = m(x - x_0) \tag{13-2}$$

Now, in the language of algebra, "x" and "y" are **variables** (may take on any value from some previously stipulated set of numbers), and "x_0" and "y_0" are **constants** (having a fixed, unchanging value). As an example, if $m = 2$ and $P_0:(3, -1)$, then the resulting expression becomes $[y - (-1)] = 2(x - 3)$, which is equivalent to $2x - y - 7 = 0$. The import of this paragraph is: The coordinates of every point on a line satisfy some algebraic equation of the first degree in two variables.

The converse of this statement is also true: All points whose coordinates satisfy some algebraic equation of the first degree in two variables lie on the same straight line.

Putting these two statements together, we can say that a line is completely characterized by an algebraic equation of the first degree in two variables.

EXAMPLES

(e) What is the equation of the line with a slope of $-\frac{5}{3}$ passing through $(2, -2)$?

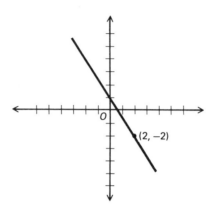

Using equation (13-2),

$$[y - (-2)] = -\frac{5}{3}(x - 2)$$

which simplifies to

$$5x + 3y - 4 = 0$$

(f) What is the equation of the line passing through $P:(-1, -4)$ and $Q:(4, 0)$?

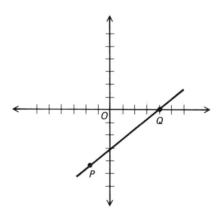

Before using Eq. (13-2), we must find the slope of the line. This can be done by resorting to Definition 13.1. Thus,

$$m = \frac{-4 - 0}{-1 - 4} = \frac{-4}{-5} = \frac{4}{5}$$

Use Eq. (13-2) with P; thus,

$$[y - (-4)] = \frac{4}{5} [x - (-1)]$$

which simplifies to $4x - 5y - 16 = 0$.   Use Eq. (13-2) with Q; thus,

$$(y - 0) = \frac{4}{5} (x - 4)$$

which also simplifies to

$$4x - 5y - 16 = 0$$

(g) What are the points on the x- and y-axis (the intercepts) where the line whose equation is $2x - y + 6 = 0$ intersects those axes?

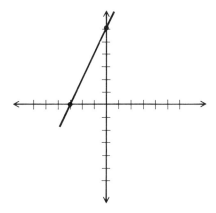

Since every point on the y-axis has an abscissa of 0, substitute 0 for x in the given equation.

$$2(0) - y + 6 = 0 \quad \text{gives } y = 6$$

Because $(0, 6)$ satisfies the equation (makes it a true statement), the y-intercept is the point $(0, 6)$.

Similarly, every point on the x-axis has an ordinate of 0.

If we substitute 0 for y, $2x - (0) + 6 = 0$ gives $x = -3$.

Because $(-3, 0)$ satisfies the equation, the x-intercept is the point $(-3, 0)$.

(h) What is the slope of the line in example (g)?

If we use the two points, $(0, 6)$ and $(-3, 0)$, and Definition 13-1,

$$m = \frac{6 - 0}{0 - (-3)} = \frac{6}{3} = 2$$

Another method is to try to put the equation in the form of Eq. (13-2). This can be considerably simplified if we choose x_0 of that equation to be 0. Then, by a rewriting of Eq. (13-2), we get $y = mx + y_0$. Thus, by "solving for y", we get an equation wherein the coefficient of x is the slope of the line and the constant term is the ordinate of the y-intercept.

For the equation of example (g):

$$2x - y + 6 = 0 \Rightarrow y = 2x + 6$$

The coefficient of x is 2; therefore, the slope is 2.

(i) Use the second method of example (h) to find the slope of the line whose equation is $3x + 5y + 2 = 0$.

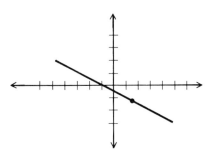

$$3x + 5y + 2 = 0 \Rightarrow 5y = -3x - 2$$

$$\Rightarrow y = -\frac{3}{5} x - \frac{2}{5}$$

The slope of the line is $-\frac{3}{5}$.

EXERCISE SET 13.2

1. Write the equations of the lines passing through the following pairs of points.

   (a) $P:(3, -4)$ and $Q:(1, -2)$ (f) $X:(-5, 0)$ and $Y:(5, -3)$

   (b) $A:(-2, -3)$ and $B:(1, 7)$ (g) $J:(1, -3)$ and $K:(1, -7)$

   (c) $M:(0, 4)$ and $N:(3, 1)$ (h) $L:(2, \frac{5}{2})$ and $M:(0, 4)$

   (d) $E:(0, -2)$ and $F:(2, 0)$ (i) $B:(1, 7)$ and $G:(-1, \frac{17}{2})$

   (e) $R:(6, 2)$ and $T:(-4, 2)$ (j) $S:(\frac{3}{4}, -\frac{3}{2})$ and $V:(\frac{27}{4}, \frac{13}{2})$

2. Which of the lines in Exercise 1 are parallel and which are perpendicular?

3. Prove that the following points are collinear: $M:(3, -2)$, $N:(4, 0)$, and $P:(7, 6)$.

4. Given the points $A:(-1, -5)$, $B:(2, -2)$, $C:(3, 1)$, and $D:(0, -2)$, show that $ABCD$ is a parallelogram.

5. Given the points $W:(3, -2)$, $X:(6, 2)$, $Y:(2, 5)$, and $Z:(-1, 1)$, prove that $WXYZ$ is a rectangle.

6. Prove that the triangle with the vertices $R:(12, -17)$, $S:(-11, 2)$, and $T:(10, 4)$ is a right triangle.

7. Give the slopes and x- and y-intercepts of the following lines:

   (a) $3x - 2y + 7 = 0$ (f) $7x + 7y + 7 = 0$

   (b) $x + 5y - 3 = 0$ (g) $x = y$

   (c) $2x + 3y + 4 = 0$ (h) $\frac{3}{4}x - \frac{4}{3}y = 24$

   (d) $6x - 4y + 5 = 0$ (i) $2y - 2x + 6 = 0$

   (e) $x + y = 0$ (j) $5x - y + 11 = 0$

8. Which of the lines in Exercise 7 are parallel and which are perpendicular?

9. For each of the lines in Exercise 7, give the coordinates of two points other than the x- and y-intercepts that lie on the line.

13.3 THE DISTANCE FORMULA

In addition to determining the equation of a line passing through two given points, the coordinates of those points enable us to calculate the length of the segment connecting those two points. This length is also called the **distance** between the two points. Regardless of the name, this information is absolutely essential if we wish to study geometry, using a coordinate system as a frame of reference. Fortunately, the formula for finding the distance

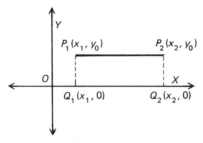

Figure 13-9

between two given points is an easy one, both to derive as well as to use. Its development is based upon the Pythagorean Theorem.

First, we shall consider the special cases where the two points determine a segment that is parallel to, or lying on, one of the coordinate axes. If the segment is parallel to the x-axis, as shown in Fig. 13-9, draw parallels to the y-axis from the endpoints of the segment, meeting the x-axis in points Q_1 and Q_2, having coordinates $(x_1, 0)$ and $(x_2, 0)$, respectively. Since $P_1P_2Q_2Q_1$ is a rectangle,

$$m(\overline{Q_1Q_2}) = m(\overline{P_1P_2})$$

But Q_1 and Q_2 lie on the x-axis, which is a coordinate line. Therefore, the length of $\overline{Q_1Q_2}$ is $|x_2 - x_1|$. This value, then, is also the length of $\overline{P_1P_2}$.

In a similar fashion, the length of a segment parallel to the y-axis may be found. If the coordinates of the endpoints are (x_0, y_1) and (x_0, y_2), then the length of the segment is $|y_2 - y_1|$. See Fig. 13-10.

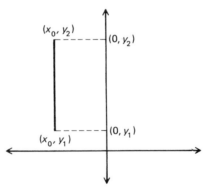

Figure 13-10

If the segment is neither horizontal nor vertical, then we proceed by drawing the lines as indicated in Fig. 13-11. When we draw the lines parallel to the coordinate axes, a right triangle is formed with right angle at

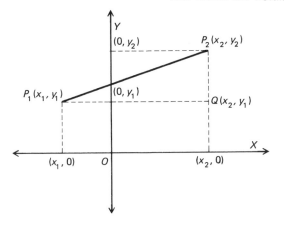

Figure 13-11

$Q.$ Using the Pythagorean Theorem with $\triangle P_1 P_2 Q$, we get

$$[m(\overline{P_1P_2})]^2 = [m(\overline{P_1Q})]^2 + [m(\overline{P_2Q})]^2$$

Let the length of $\overline{P_1P_2}$ be symbolized by $m(\overline{P_1P_2})$. Using the ideas from the previous two paragraphs, and noting that

$$|x_2 - x_1|^2 = (x_2 - x_1)^2$$

and similarly for the "y's", we derive the following formula:

THEOREM 13.4 (The Distance Formula)

The distance between two points, $P_1:(x_1, y_1)$ and $P_2: (x_2, y_2)$ may be found by the following formula:

$$m(\overline{P_1P_2}) = \sqrt{(x_2 - x_1)^2 + (y_2 - y_1)^2}$$

EXAMPLE

Find the distance between $A:(4, -1)$ and $B:(-2, 7)$.

$$m(\overline{AB}) = \sqrt{[4 - (-2)]^2 + (-1 - 7)^2} = \sqrt{(6)^2 + (-8)^2}$$

$$= \sqrt{36 + 64}$$

$$= \sqrt{100}$$

$$= 10$$

Thus, $m(\overline{AB}) = 10$.

Note that the distance formula also includes the special cases of the vertical and horizontal segments.

With the distance formula, we may now complete the proof of Theorem 13.3.

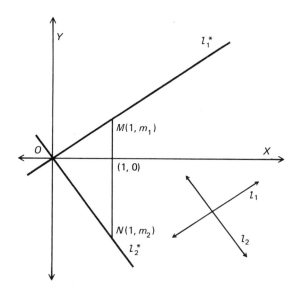

Figure 13-12

Proof of Theorem 13.3.   Refer to Fig. 13-12.   If necessary, draw l_1^* $\parallel$ l_1 so that l_1^* passes through the origin.   In a like fashion, construct l_2^* through O so that l_2^* $\parallel$ l_2. By Theorem 13.2, l_1^* and l_2^* have slopes m_1 and m_2, respectively.   Thus, $l_1 \perp l_2$ if and only if $l_1^* \perp l_2^*$.   At a point $(1, 0)$, erect a perpendicular to the x-axis, meeting l_1^* at M and l_2^* at N.   The x-coordinates of M and N are 1.   To find the y-coordinates, we do as follows.   Let M and O be the two points of l_1^* that give us the slope, which is known to be m_1.   Thus,

$$m_1 = \frac{y_m - 0}{1 - 0}$$

Therefore, y_m, the y-coordinate of M, equals m_1.   In a similar fashion, the y-coordinate of N can be shown to be equal to m_2.   The remainder of the proof will use the Pythagorean Theorem and its converse with $\triangle MON$. This says that $\triangle MON$ is a right triangle with right angle at O if and only if $[m(\overline{MN})]^2 = [m(\overline{MO})]^2 + [m(\overline{NO})]^2$.   Using the Distance Formula and some algebraic simplification, we achieve the desired result.   All this is shown symbolically in the following lines.

$$l_1 \perp l_2 \; l_1^{*}\perp \Leftrightarrow l_2^{*} \Leftrightarrow [m(\overline{MN})]^2 = [m(\overline{MO})]^2 + [m(\overline{NO})]^2$$

(Pythagorean Theorem)

$$\Leftrightarrow [(m_2 - m_1)^2 + (1 - 1)^2] =$$

$$[(m_1 - 0)^2 + (1 - 0)^2] + [(m_2 - 0)^2 + (1 - 0)^2]$$

(Distance Formula)

$$\Leftrightarrow m_2^2 - 2m_1 m_2 + m_1^2 = m_2^2 + 1 + m_1^2 + 1$$

$$\Leftrightarrow -2m_2 m_1 = 2$$

$$\Leftrightarrow -m_2 m_1 = 1$$

$$\Leftrightarrow m_1 = -\frac{1}{m_2}$$

The continued use of logical equivalence ("⇔") at each step shows that both the "if" and "only if" parts of the theorem have been proven at the same time.

The Distance Formula also has an added "bonus," in that it enables us to write the equation of a circle. A circle is a set of all points that are at a fixed distance from a fixed point. Rather than symbolize this distance by "d," let us use the familiar r for the radius. Let the coordinates of the fixed point be (h, k)—these two letters, "h" and "k," are used in practically all textbooks as the coordinates of the center of a circle—and let $P:(x, y)$ be the representation of all the points of the circle. Squaring both sides of the distance formula gives the form in which the equation is normally written.

THEOREM 13.5 (Equation of a Circle)

The equation of a circle containing all points $P:(x, y)$ at a distance r from a fixed point $C:(h, k)$ is:

$$(x - h)^2 + (y - k)^2 = r^2$$

EXAMPLE

What is the equation of the circle with center $(1, -3)$ and radius of 2?

If we use Theorem 13.5,

$$(x - 1)^2 + (y + 3)^2 = 4$$

13.4 THE MIDPOINT FORMULA

It is often convenient to be able to express the coordinates of the midpoint of a segment in terms of the coordinates of the endpoints of the segment. The derivation of the midpoint formula depends upon one theorem, whose statement and proof have not been given. It states that if a line is drawn through the midpoint of one side of a triangle parallel to a second side, then it bisects the third side. This theorem is illustrated in Fig. 13-13. Line m is parallel to $\overline{AC}$. If D is the midpoint of $\overline{AB}$, then E is the midpoint of $\overline{BC}$.

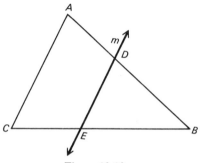

Figure 13-13

Applying this fact to coordinate geometry, we can find the coordinates of the midpoint of a given segment. Let $P_1:(x_1, y_1)$ and $P_2:(x_2, y_2)$ be the endpoints of a segment and $P:(x, y)$ be the midpoint of that segment. Drawing parallels to the x- and y-axes through P_1, P_2, and P gives the configuration illustrated in Fig. 13-14. $R:(x_2, y)$ is the midpoint of $\overline{P_2Q}$ and $S:(x, y_1)$ is the midpoint of $\overline{P_1Q}$.

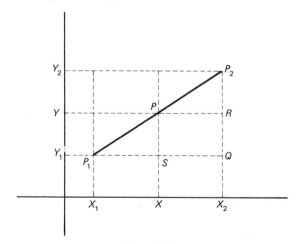

Figure 13-14

292

These facts tell us that

$$m(\overline{Y_1Y}) = m(\overline{YY_2}) \quad \text{and} \quad m(\overline{X_1X}) = m(\overline{XX_2})$$

If these latter equations are expressed in coordinate form, the desired results appear readily.

$$m(\overline{Y_1Y}) = m(\overline{YY_2}) \Rightarrow (y - y_1) = (y_2 - y)$$

$$\Rightarrow 2y = y_1 + y_2$$

$$\Rightarrow y = \frac{y_1 + y_2}{2}$$

$$m(\overline{X_1X}) = m(\overline{XX_2}) \Rightarrow (x - x_1) = (x_2 - x)$$

$$\Rightarrow 2x = x_1 + x_2$$

$$\Rightarrow x = \frac{x_1 + x_2}{2}$$

We have proven the following theorem.

THEOREM 13.6 (Midpoint Formula)

If $P_1 : (x_1, y_1)$ and $P_2 : (x_2, y_2)$ are the endpoints of a segment, the coordinates of the midpoint $P : (x, y)$ of that segment are given by:

$$x = \frac{x_1 + x_2}{2} \quad \text{and} \quad y \quad \frac{y_1 + y_2}{2}$$

EXAMPLE

What are the coordinates of the midpoint of $\overline{AB}$ if $A : (3, -7)$ and $B : (-6, -11)$?

$$x = \frac{3 + (-6)}{2} = \frac{-3}{2} = -\frac{3}{2} \qquad y = \frac{-7 + (-11)}{2} = \frac{-18}{2} = -9$$

Therefore, the midpoint is $(-\frac{3}{2}, -9)$.

EXERCISE SET 13.3

1. Find the distance between the named points of Exercise 1 of Exercise Set 13.2.

2. Find the coordinates of the midpoints of the segments determined by the named points of Exercise 1 of Exercise Set 13.2.

3. Find the perimeter of the following triangles:

(a) $A:(5, -2)$, $B:(7, -6)$, $C:(1, 0)$
(b) $M:(0, 4)$, $N:(-3, 2)$, $P:(1, -5)$
(c) $X:(1, 1)$, $Y:(5, 4)$, $Z:(-2, -3)$
(d) $O:(0, 0)$, $R:(-3, -6)$ $S:(-3, 6)$

4. Which of the triangles in Exercise 3 are isosceles? Which are equilateral?

5. Given the points $A:(-3, 6)$, $B:(5, 7)$, $C:(9, 0)$, and $D:(1, -1)$, prove $ABCD$ is a rhombus.

6. For the points $M:(-2, -9)$, $N:(1, -5)$, and $Q:(5, -2)$, prove that N is not between M and Q.

7. If the vertices of a triangle are $R:(4, 1)$, $S:(0, 7)$, and $T:(-4, 3)$, what are the lengths of its three medians?

8. The coordinates of the vertices of a quadrilateral are $A:(-1, -3)$, $B:(1, -2)$, $C:(4, 3)$, and $D:(-1, 1)$. Show that the coordinates of the midpoints of the sides of this quadrilateral are the vertices of a parallelogram.

9. What is the equation of the circle with center at the origin and a radius of r?

10. Write the equations of the following circles:

	Center	Radius
(a)	$(0, -5)$	1
(b)	$(1, 4)$	5
(c)	$(-2, 0)$	2
(d)	$(-1, -3)$	2
(e)	$(2, -1)$	3

11. Does the line $2x + y - 7 = 0$ intersect the circle $(x + 2)^2 + (y - 3)^2 = 9$? How can you tell without drawing the graphs of the two?

13.5 PROOFS OF SOME THEOREMS BY MEANS OF COORDINATE METHODS

Many of the theorems of plane geometry can be proven quite easily by means of the formulas developed in the previous sections. It is the purpose of this section to exemplify some of these proofs, not to make the reader an expert in this approach, but rather to demonstrate the power of the method.

The value of the coordinate method depends upon the simplicity of the algebra involved. The algebraic manipulations can be made easy by choosing the appropriate coordinate system. The plane can be coordinated in infinitely many ways; thus, we are free to choose a way that will make the work tolerable. Ordinarily, we do this by choosing the coordinate axes so that at least one of the axes will coincide with a side of a polygon and so that vertices lie on the axes wherever possible. For example, if we were proving a theorem about a rectangle, the configuration of part (b) of Fig. 13-15 would be much more desirable than that in part (a).

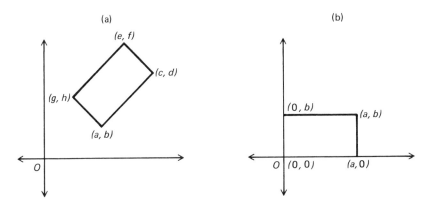

Figure 13-15

On the other hand, we must be careful not to impose a condition upon a geometric figure that may not be imposed. For example, if a general triangle were the subject of interest, it would be incorrect to place the vertices of the triangle at the origin, at a point of the x-axis, and at a point of the y-axis. To do so would make the triangle a right triangle; as such, it possesses properties that not all triangles have.

With these remarks in mind, let us examine several examples of co-ordinate proofs.

EXAMPLES

(a) The diagonals of a rectangle are congruent.

The coordinates of the plane are established, so that the rectangle is placed as shown in the figure.   The Distance Formula is utilized.

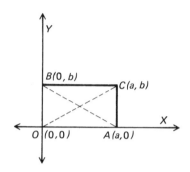

$$m(\overline{OC}) = \sqrt{(a-0)^2 + (b-0)^2} = \sqrt{a^2 + b^2}$$

$$m(\overline{AB}) = \sqrt{(0-a)^2 + (b-0)^2} = \sqrt{a^2 + b^2}$$

Thus, the two diagonals are congruent since their lengths are equal.

(b) The midpoint of the hypotenuse of a right triangle is equidistant from the vertices.

The triangle is pictured below.   In proving theorems involving midpoints of segments, we find it convenient to give the endpoints coordinates such as $(2x, 2y)$, so that the coordinates of the midpoints do not have fractions. In this theorem, point D is, by definition, equidistant from A and

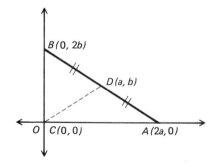

B. Thus, all that is needed is to show

$$m(\overline{CD}) = m(\overline{AD})$$

$$m(\overline{CD}) = \sqrt{(a-0)^2 + (b-0)^2} = \sqrt{a^2 + b^2}$$

$$m(\overline{AD}) = \sqrt{(a-2a)^2 + (b-0)^2} = \sqrt{a^2 + b^2}$$

The theorem is proved.

(c) The segment connecting the midpoints of the nonparallel sides of a trapezoid is parallel to the bases and has a measure equal to one-half the sum of the measures of the bases. We again place the trapezoid appropriately. Since the bases are defined to be parallel and since the lower base lies on the *x*-axis, the upper base, $\overline{BC}$, is a horizontal segment. Therefore, the *y*-coordinates of *B* and *C* are equal. The coordinates of *M* and *N* result from the Midpoint Formula.

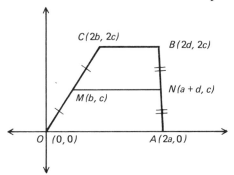

(1) To show $\overline{MN} \parallel \overline{AO} \parallel \overline{BC}$. Since $\overline{AO}$ and $\overline{BC}$ are horizontal, their slopes are equal to zero. We must show $m_{MN} = 0$. But

$$m_{MN} = \frac{c - c}{(a+d) - b} = \frac{0}{a+d-b} = 0$$

Therefore,

$$\overline{MN} \parallel \overline{AO} \parallel \overline{BC}$$

(2) To show $m(\overline{MN}) = \frac{1}{2}[m(\overline{AO}) + m(\overline{BC})]$. Solving this part is aided considerably by the fact that all three segments are horizontal.

$$m(\overline{MN}) = a + d - b$$

$$m(\overline{AO}) = 2a$$

$$m(\overline{BC}) = 2d - 2b$$

$$\tfrac{1}{2}\left[m(\overline{AO}) + m(\overline{BC})\right] = \tfrac{1}{2}[2a + (2d - 2b)]$$

$$= a + d - b = m(\overline{MN})$$

Thus, the second part of the theorem is proven.

(d) The line segment connecting the midpoints of two sides of a triangle is parallel to the third side and has a measure equal to one-half the measure of the third side.

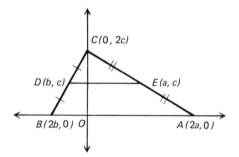

The proof of these two parts is identical to the two parts of example (c). This example has been given for the purpose of demonstrating the placement of the triangle.

(e) The diagonals of a rhombus are perpendicular.
The labeling of points A and C is as before. Point B has the labeled coordinates for the following reasons.

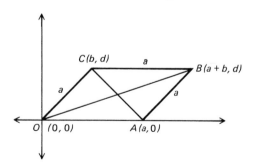

(1) The y-coordinate.   A rhombus is a parallelogram; thus $\overline{CB} \parallel \overline{AO}$. But, $m_{AO} = 0$; $\overline{AO}$ is horizontal. Thus, $\overline{CB}$ is horizontal, and the y-coordinates of its endpoints must be equal.

(2) The x-coordinate.   A rhombus has all four sides congruent. Since $m(\overline{OA}) = a$, $m(\overline{CB}) = a$. But, $m(\overline{CB}) = x_B - b$. Thus,

$$x_B - b = a \quad \text{or} \quad x_B = a + b$$

$$m(\overline{OC}) = \sqrt{(b - 0)^2 + (d - 0)^2} = \sqrt{b^2 + d^2} = a$$

Thus,

$$b^2 + d^2 = a^2 \tag{13-3}$$

$$m_{AC} = \frac{d - 0}{b - a}, \qquad m_{OB} = \frac{d - 0}{a + b - 0} = \frac{d}{a + b}$$

$$m_{AC} \cdot m_{OB} = \frac{d}{b - a} \cdot \frac{d}{a + b} = \frac{d^2}{b^2 - a^2}$$

From Eq. (13-3),

$$d^2 = a^2 - b^2$$

Thus,

$$m_{AC} \cdot m_{OB} = \frac{a^2 - b^2}{b^2 - a^2} = -1$$

Therefore,

$$m_{AC} = \frac{-1}{m_{OB}}$$

which, by Theorem 13.3, tells us $\overline{AC} \perp \overline{OB}$.

(f) The altitudes of a triangle are concurrent.

Although the figure shows O between A and B, the following proof is identical for A and B, both on the positive or both on the negative x-axis. The y-axis serves as the altitude from C to $\overline{AB}$. Thus, to show all three altitudes concurrent, we must demonstrate that the altitudes from B to $\overline{AC}$ and from A to $\overline{BC}$ have the same y-intercepts.

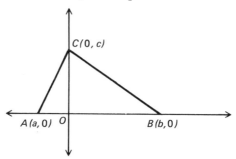

This may be done by examining the equations of these lines.

$$m_{AC} = \frac{c - 0}{0 - a} = -\frac{c}{a}$$

Thus, the slope of the altitude to $\overline{AC}$ is a/c.

$$m_{BC} = \frac{c - 0}{0 - b} = -\frac{c}{b}$$

Thus, the slope of the altitude to $\overline{BC}$ is b/c.

The equation of the altitude to $\overline{AC}$ is

$$(y - 0) = \frac{a}{c}(x - b) \quad \text{or} \quad y = \frac{a}{c}x - \frac{ab}{c}$$

The equation of the altitude to $\overline{BC}$ is

$$(y - 0) = \frac{b}{c}(x - a) \quad \text{or} \quad y = \frac{b}{c}x - \frac{ab}{c}$$

Both of these equations are in the form where the y-intercept may be read directly from the equation. In both instances, it is the same:

$$\left(0, -\frac{ab}{c}\right)$$

Therefore, the altitudes are concurrent.

SUGGESTED READINGS

[9] Ch. 10

[10] Ch. 2

[21] Ch. 9

BIBLIOGRAPHY

1. Anderson, Richard D., Jack W. Garon, and Joseph G. Gremillion, *School Mathematics Geometry*. Boston: Houghton Mifflin Company, 1966.
2. Arnold, B. H., *Intuitive Concepts in Elementary Topology*. Englewood Cliffs, N. J.: Prentice-Hall, Inc., 1962.
3. Barr, Stephen, *Experiments in Topology*. New York: Thomas Y. Crowell Company, 1964.
4. Barry, Edward H., *Introduction to Geometrical Transformations*. Boston: Prindle, Weber and Schmidt, Inc., 1966.
5. Chinn, W. G., and N. E. Steenrod, *First Concepts of Topology*. New York: The L. W. Singer Company, 1966.
6. Dorwart, Harold L., *The Geometry of Incidence*. Englewood Cliffs, N. J.: Prentice-Hall, Inc., 1966.
7. Eves, Howard, *A Survey of Geometry, Vols. I & II*. Boston: Allyn & Bacon, Inc., 1963.
8. Garstens, Helen L., and Stanley B. Jackson, *Mathematics for Elementary School Teachers*. New York: The Macmillan Company, 1967.
9. Keedy, Mervin L., and Charles W. Nelson, *Geometry: A Modern Introduction*. Reading, Mass.: Addison-Wesley Publishing Company, Inc., 1965.
10. Kelly, Paul, and Norman E. Ladd, *Analytic Geometry*. Glenwood, Ill.: Scott, Foresman & Company, 1968.
11. ———, *Geometry*. Chicago: Scott, Foresman & Company, 1965.
12. Kostovskii, A. N., *Geometrical Constructions Using Compasses Only*. (Translated by Halina Moss) New York: Blaisdell Publishing Company, 1961.
13. Moise, Edwin E., *Elementary Geometry from an Advanced Viewpoint*. Reading, Mass.: Addison-Wesley Publishing Company, 1963.

14. National Council of Teachers of Mathematics, *The Arithmetic Teacher.* See especially the issues of Jan. 1966, May 1967, Oct. 1967, Oct. 1969.

15. ———, *Insights into Modern Mathematics.* Washington, D. C.: National Council of Teachers of Mathematics, 1957.

16. Newman, James R. (ed.), *The World of Mathematics.* New York: Simon and Schuster, Inc., 1956.

17. Norton, M. Scott, *Geometric Constructions.* St. Louis: Webster Division, McGraw-Hill Book Company, 1963.

18. Prenowitz, Walter, and Meyer Jordan, *Basic Concepts of Geometry.* New York: Blaisdell Publishing Company, 1965.

19. Rainich, G. Y., and S. M. Dowdy, *Geometry for Teachers.* New York: John Wiley & Sons, Inc., 1968.

20. Ringenberg, Lawrence A., *Informal Geometry.* New York: John Wiley & Sons, Inc., 1967.

21. ———, *College Geometry.* New York: John Wiley & Sons, Inc., 1968.

22. Rosskopf, Myron F., Joan L. Levine, and Bruce R. Vogeli, *Geometry, A Perspective View.* New York: McGraw-Hill Book Company, 1969.

23. School Mathematics Study Group, *Studies in Mathematics, Vol. IV—Geometry.* SMSG, 1961.

24. ———, *Studies in Mathematics, Vol. V—Concepts of Informal Geometry.* SMSG, 1961.

25. ———, *Studies in Mathematics, Vol. VII—Intuitive Geometry.* SMSG, 1961.

26. Smart, James R., *Introductory Geometry: An Informal Approach.* Belmont, Calif.: Brooks/Cole Publishing Company, Inc., 1967.

27. Spooner, George A., and Richard L. Mentzer, *Introduction to Number Systems.* Englewood Cliffs, N. J.: Prentice-Hall, Inc., 1968.

28. Steiner, Jacob, *Geometrical Constructions with a Ruler.* (Translated by Marion Stark) New York: Scripta Mathematica, 1950.

29. Wylie, C. R., Jr., *Foundations of Geometry.* New York: McGraw-Hill Book Company, 1964.

ANSWERS TO
SELECTED EXERCISES

Exercise Set 2.1

1. (a) Yes. (b) No. (c) Yes. (d) Yes. (e) No. (f) Yes.

2. (a) $\{a\}$, $\varnothing$.

   (b) $\{\triangle\}$, $\{\square\}$, $\{\triangle,\square\}$, $\varnothing$.

   (c) $\{m\}$, $\{n\}$, $\{p\}$, $\{m,n\}$, $\{m,p\}$, $\{n,p\}$, $\{m,n,p\}$, $\varnothing$.

   (d) $\{w\}$ $\{x\}$, $\{y\}$, $\{z\}$, $\{w,x\}$, $\{w,y\}$, $\{w,z\}$, $\{x,y\}$,
     $\{x,z,\}$, $\{y,z\}$, $\{w,x,y\}$, $\{w,x,z\}$, $\{w,y,z\}$, $\{x,y,z\}$,
     $\{w,x,y,z\}$, $\varnothing$.

3. (a) A: 2; F: 4; T: 8; B: 16.

   (b) Five elements: 32; six elements: 64.

4. 2^n.

Exercise Set 2.2

1. (a) $\varnothing$. (b) A. (c) A. (d) A. (e) B. (f) A. (g) A. (h) B.

2. (a) $\{c\}$. (b) $\{X, Y\}$. (c) $\varnothing$. (d) M. (e) Y. (f) $\{d\}$. (g) $\varnothing$.

3. (a) $\{a, b, c, d, e\}$.   (e) X.
   (b) $\{W, X, Y, Z\}$.   (f) $\{a, b, c, d, e, f\}$.
   (c) $\{0, 1, 2, 3, 4, 5, 6\}$.   (g) $\{2, 3, 4, 5, 6, 7, 9\}$.
   (d) M.

Exercise Set 2.3

1. (a) $\overline{NP}$. (b) $\overline{PQ}$. (c) $\overrightarrow{NR}$. (d) $\varnothing$. (e) P. (f) $\varnothing$. (g) $\overrightarrow{MQ}$. (h) $\overleftrightarrow{MR}$.

2. (a) $\overline{AB} \cap \overline{BC}$ or $\overline{AB} \cap \overline{BD}$ or $\overline{AC} \cap \overline{CD}$.
   (b) $\overline{AB}$ and $\overline{CD}$.
   (c) There are ten such pairs. One of them is $\overline{AD} \cap \overline{BD}$.
   (d) There are twelve such pairs. One of them is $\overline{AC} \cup \overline{AD}$.
   (e) $\overline{AB} \cup \overline{CD}$.

3. (a) $\overrightarrow{BA} \cap \overrightarrow{BC}$ or $\overrightarrow{BA} \cap \overrightarrow{BD}$ or $\overrightarrow{CA} \cap \overrightarrow{CD}$.
   (b) $\overrightarrow{BA}$ and $\overrightarrow{CD}$.
   (c) $\overrightarrow{BA} \cap \overrightarrow{AB}$; $\overrightarrow{BC} \cap \overrightarrow{CB}$; $\overrightarrow{CD} \cap \overrightarrow{DC}$; $\overrightarrow{BC} \cap \overrightarrow{DC}$; $\overrightarrow{CA} \cap \overrightarrow{AB}$;
   $\overrightarrow{CB} \cap \overrightarrow{AB}$; $\overrightarrow{DC} \cap \overrightarrow{AB}$.
   (d) There are no such pairs.
   (e) Any two rays will satisfy this condition.

4.
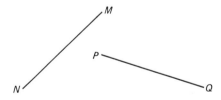

5. (a) 3: $\overline{XY}$, $\overline{XZ}$, $\overline{YZ}$.
   (b) 6: The names depend upon the relative placement of points.
   (c) 6: The names depend upon the relative placement of points.

6. (a) 6: $\overline{XY}$, $\overline{XZ}$, $\overline{YZ}$, $\overline{XW}$, $\overline{YW}$, $\overline{ZW}$.
   (b) 8: The names depend upon the relative placement of points.
   (c) 8: The names depend upon the relative placement of points.

7. *Hint*: Consider the following table:

No. of points	0	1	2	3	4	5
No. of segments	0	0	1	3	6	—
No. of rays	0	2	4	6	8	—
No. of half-lines	0	2	4	6	8	—

8. 3: $\overline{DE}, \overline{EF}, \overline{DF}$.

9. 6: $\overline{DE}, \overline{DF}, \overline{EF}, \overline{DG}, \overline{EG}, \overline{FG}$.

10. *Hint*: Consider the first two rows of the table in answer to Exercise 7 of this set.

11. A ray contains an endpoint; the half-line does not.

12. $\overrightarrow{BA} = \overline{BA} \cup \{$ all points C on $\overleftrightarrow{BA}$ such that A is between B and $C\}$.

Exercise Set 2.4

1. (a) Yes. (b) Yes. (c) Yes. (d) Yes. (e) Yes. (f) Yes. (g) No. (h) No. (i) Yes. (j) Yes. (k) No. (l) Yes.

2. Call the points P and Q. If $\overline{PQ}$ intersects the boundary line l, P and Q are not in the same half-plane; if $\overline{PQ}$ does not intersect l, then P and Q are in the same half-plane.

3. 6; four elements: 24; five elements: 120; n elements: $n!$; no one-to-one correspondence; no one-to-one correspondence.

6.

Set	Binary relation?	Equivalence relation?
(a)	Yes	No
(b)	No	No
(c)	Yes	No
(d)	Yes	Yes
(e)	Yes	Yes
(f)	Yes	Yes
(g)	Yes	No
(h)	Yes	Yes

9. No.

10. $a > c$; "is greater than" is a transitive relation.

11. No.

13. Infinitely many.

Exercise Set 4.3

1. (a) $2; -3; 4; 5; -1; -2$.
   (b) $1; 5; 3; 2; 2; 6$.

2. (a) $2; -\frac{1}{2}; 3; \frac{7}{2}; \frac{1}{2}; 0$.
   (b) $\frac{1}{2}; \frac{5}{2}; \frac{3}{2}; 1; 1; 3$.

3. (a) $0; \frac{5}{3}; -\frac{2}{3}; -1; 1; \frac{4}{3}$.
   (b) $\frac{1}{3}; \frac{5}{3}; 1; \frac{2}{3}; \frac{2}{3}; 2$.

4. $A = B$.

6. B is between A and C if $a > b > c$ or $a < b < c$.

8. (a) $=$ (d) $<$ (g) $<$ (j) $=$
   (b) $<$ (e) $=$ (h) $=$
   (c) ?   (f) $=$ (i) $<$

Exercise Set 4.4

1. (a) C: 60 S: 150.   (d) C: $210 - x$ S: $300 - x$.
   (b) C: $42\frac{1}{2}$ · S: 132.   (e) C: $90 - 20\sqrt{2} + 5x$.
   (c) C: 55.1 S: 145.1. S: $180 - 20\sqrt{2} + 5x$.

2. (d) $120 < x < 210$.   (e) $4\sqrt{2} - 18 < x < 4\sqrt{2}$.
       $120 < x < 300$. $4\sqrt{2} - 36 < x < 4\sqrt{2}$.

4. It should be close to 180.

5. No.

6. (a) 60-acute   (c) 130-obtuse   (e) 90-right   (g) 145-obtuse
   (b) 80-acute   (d) 85-acute   (f) 60-acute   (h) 145-obtuse

7. Yes.

8. Yes.

9. They are congruent. Yes.

(i)	No	No
(j)	Yes	No
(k)	Yes	Yes
(l)	No	No

7. Four; seven; eleven.

8. Four; eight; fifteen; twenty-six.

Exercise Set 3.1

7. No.

9.

No points	1 point	2 points	3 po

Exercise Set 3.2

2. They are collinear.

Exercise Set 3.3

1. Not counting the zero or straight angle, 5. $\angle APB$, $\angle APC$, $\angle BPC$, $\angle BPL$ $\angle CPD$. 4; $\angle APB$ and $\angle BPC$; $\angle APB$ and $\angle BPD$; $\angle BPC$ and $\angle CPD$ $\angle APC$ and $\angle CPD$.

2. 2; $\angle APB$ and $\angle BPD$; $\angle APC$ and $\angle CPD$.

Exercise Set 4.2

1. Addition: No, e.g., $2 + 1 = 3 \notin A$.
   Multiplication: No, e.g., $2 \times 2 = 4 \notin A$.

4. No. Only transitivity holds.

5. (a) $x > 3$. (d) $x > 9$. (g) $x \le -\frac{11}{5}$. (j) $-2 < x < 20$.
   (b) $x < 34$. (e) $x \le \frac{13}{5}$. (h) $x > -7$. (k) $-\frac{4}{3} < x < \frac{2}{3}$.
   (c) $x \ge 6$. (f) $x < 2$. (i) $1 > x \ge -\frac{3}{2}$

Exercise Set 4.5

1.

Measurement	Precision Unit	g.p.e.	Relative Error
$3\frac{5}{8}$ in.	$\frac{1}{8}$ in.	$\frac{1}{16}$ in.	$\frac{1}{58} = 1.7\%$
3.060 ft	0.001 ft	0.0005 ft	$\frac{1}{6120} = 0.016\%$
2400 m	100 m	50 m	$\frac{1}{48} = 2.08\%$
240.0 m	0.1 m	0.05 m	$\frac{1}{4800} = 0.0208\%$
0.003 mm	0.001 mm	0.0005 mm	$\frac{1}{6} = 16.7\%$

2. (a) $7\frac{1}{4}$ to $7\frac{3}{4}$ in.
   (b) $2\frac{1}{32}$ to $2\frac{3}{32}$ yds.
   (c) $4\frac{1}{8}$ to $4\frac{3}{8}$ ft.
   (d) 6.825 to 6.835 in.
   (e) 2.55 to 2.65 dm.
   (f) 1.0295 to 1.0305 mm.
   (g) 5.5 to 6.5 km.
   (h) 2.425 to 2.435 cm.

3. Exact.

4. (a) $\frac{1}{4}$ in.
   (b) $\frac{1}{32}$ yd.
   (c) $\frac{1}{8}$ ft.
   (d) 0.005 in.
   (e) 0.05 dm.
   (f) 0.0005 mm.
   (g) 0.5 km.
   (h) 0.005 cm.

5. (a) 3.33%.
   (b) 1.51%.
   (c) 2.94%.
   (d) 0.073%.
   (e) 1.92%.
   (f) 0.048%.
   (g) 8.32%.
   (h) 0.23%.

6. (a) 56.
   (b) 3.
   (c) 9,407.
   (d) 2,003.
   (e) 205.
   (f) 1,470.
   (g) 403.
   (h) 268,000.

7. (a) 26,030 cm.
   (b) 0.0003 dm.
   (c) 14,600 m.
   (d) 0.0000667 dkm.
   (e) 2,600,000 cm.
   (f) 58,340 dm.
   (g) 193.45 m.
   (h) 140 m.

10. 0.01 in.

Exercise Set 5.1

1. Curves: a, b, d, e, g, i, l.
   Closed curves: d, g, i, l.
   Simple closed curves: g, i, l.

2. No; yes; yes.

4. (a) Yes. (b) No. (c) No. (d) Yes. (e) Yes. (f) Yes. (g) Yes.
   (h) Yes.

5. (a) Scalene. (d) Isosceles. (e) Scalene. (f) Equilateral. (g) Scalene.
   (h) Isosceles.

6. All angles should have equal measures.

7. The two angles opposite the congruent sides should have equal measures.

8. The angle with the largest measure is opposite the side with the largest
   measure.

9. No.

Exercise Set 5.2

1. (a) $\overline{CD}$; $\overline{CE}$; $\angle ABC$; $\angle CAB$.

   (b) $\overline{TS}$; $\overline{SR}$; $\angle RST$; $\angle MNQ$.

   (c) $\overline{AD}$; $\overline{AC}$; $\angle D$; $\angle CAB$.

   (d) $\overline{ZX}$; $\overline{WU}$; $\angle YUW$; $\angle ZXU$.

2. Yes.

3. Yes.

4. Yes.

Exercise Set 5.3

6. (a) $\overline{JK}$ and $\overline{KL}$.   (d) $\overline{DE}$ and $\overline{EF}$ (g) Depends upon the
   (b) $\angle LJK$.   (e) $\overline{DC}$ and $\overline{CE}$ triangle selected.
   (c) $\angle POQ$.   (f) $\angle XWY$ and
                                          $\angle XYW$

7. (a) SAS: $ACB \leftrightarrow LKJ$.   (d) SSS: $DEF \leftrightarrow DEC$.
   (b) ASA: $MNR \leftrightarrow PNR$.

Exercise Set 5.4

1. 1. Hypothesized as true.
2. Isosceles Triangle Theorem.
5. Definition of isosceles triangle.
6. SAS.
7. Definition of triangle congruence.
8. Definition of isosceles triangle.

3. 1. Hypothesized as true.

   3. Hypothesized as true.

   4. $\overline{BN}$ and $\overline{CM}$ are medians; definition of midpoint of a segment.

   5. Multiplication property of equality on equation of Statement 2.

   7. Definition of segment congruence.

   8. Isosceles Triangle Theorem.

   9. Reflexive property of segment congruence.

   10. SAS.

   11. Definition of triangle congruence.

Exercise Set 5.5

1. Yes.
2. 3, 4, 5, 6, n.

Exercise Set 6.1

1. (a) $\frac{16}{3}$, 8 and $\frac{9}{2}$, $\frac{33}{2}$, 3.
   (b) $\frac{4}{5}$, 3 and $\frac{1}{5}$, $\frac{7}{8}$.
   (c) 10, 2.8 and 4.5, 9.75.

2. (a) 4. (b) $\frac{32}{9}$. (c) $\pm 2\sqrt{5}$. (d) -3.

Exercise Set 6.2

1. (a) $ACB \leftrightarrow DCE$-AA.
   (b) $PQR \leftrightarrow PST$-SAS.
   (c) $PKL \leftrightarrow PMN$-AA.
   (d) $ACB \leftrightarrow ADC \leftrightarrow CDB$-AA.
   (e) $ABC \leftrightarrow NMQ$-SSS.

   (f) $TVW \leftrightarrow TXY$-AA.
   (g) $GHJ \leftrightarrow QPR$-SAS.
   (h) $ABD \leftrightarrow BCD$-AA.
   (i) $PYH \leftrightarrow AGT$-SSS.
   (j) $MKH \leftrightarrow JKW$-SAS.

2. 1. Hypothesized as true.

   3. Hypothesized as true.

   4. Definition of altitude.

   5. All right angles are congruent.

   6. AA Theorem.

   8. $\triangle BAC \sim \triangle NMQ$.

   9. Transitivity property of equality.

Exercise Set 6.3

1. (a) $\sqrt{89}$.   (c) $6\sqrt{3}$.   (e) $\sqrt{3}$.
    (b) $2\sqrt{2}$.   (d) 8.5.   (f) $2\sqrt{5}$.
2. (a) 30 ft.   (c) $\frac{57}{2}$ ft.
    (b) $\frac{144}{5}$ ft.   (d) 63.36 ft.
3. (a) 420 ft.   (c) 1201 ft.
    (b) 240 ft.   (d) 216 ft.
4. (a) 9 ft.   (c) 20 ft.
    (b) 10 ft.   (d) 24.11 ft.

Exercise Set 6.4

1. All answers are rounded off to the nearest hundredth.
    (a) 7.42 yd.   (e) 40.00 in.   (i) 12.74 ft.
    (b) 7.42 in.   (f) 63.55 yd.   (j) 3.08 cm.
    (c) 8.99 cm.   (g) 3.05 m.   (k) 66.67 mm.
    (d) 7.62 ft.   (h) 4.42 in.   (l) 2.13 m.

2. (a) 30°.   (c) 30°.   (e) 68°.
    (b) 47°.   (d) 56°.   (f) 52°.

3. Approximately 18.13 ft from the house;
    approximately 8.45 ft from the ground.

4. Approximately 9,525 ft.

5. Approximately 7,781 ft.

Exercise Set 7.2

1. (a) The number 2 is not a prime number.
    (b) Robert is a straight-A student.
    (c) California is not the most populous state in the union.
    (d) Some athletes are not overweight.
    (e) All teachers are reasonable.
    (f) Someone pays attention to me.
    (g) Nobody up there likes me.

3.

$p \wedge q$	$(p \wedge q) \wedge r$
T	T
T	F

All other entries are F in both columns.

5.

p	$\sim p$	$p \wedge \sim p$
T	F	F
F	T	F

6.

$\sim (p \wedge q)$
F
T
T
T

7.

p	q	$\sim p$	$\sim q$	$\sim p \wedge \sim q$
T	T	F	F	F
T	F	F	T	F
F	T	T	F	F
F	F	T	T	T

8.

p	q	$\sim q$	$\sim p \wedge q$
T	T	F	F
T	F	F	F
F	T	T	T
F	F	T	F

Exercise Set 7.3

4.

p	$\sim p$	$p \vee \sim p$
T	F	T
F	T	T

5.

$$(p \lor q)$$

F
F
F
T

6.

p	q	$\sim p$	$\sim q$	$\sim p \lor \sim q$
T	T	F	F	F
T	F	F	T	T
F	T	T	F	T
F	F	T	T	T

7.

p	q	$\sim q$	$p \lor \sim q$	$\sim (p \lor \sim q)$
T	T	F	T	F
T	F	T	T	F
F	T	F	F	T
F	F	T	T	F

Exercise Set 7.4

1. (a) If a person is an athlete, then he is overweight.

   (b) If a person is a Martian, then he is from outer space.

   (c) If an animal is a horse, then it has four legs.

   (d) If a person is a freshman, then he is a serious student.

   (e) If I eat tamales, then I get heartburn.

   (f) If a figure is a square, then it is a rectangle.

3.

$$q \Rightarrow p$$

T
T
F
T

4.

$\sim p$	$\sim q$	$\sim p \Rightarrow \sim q$
F	F	T
F	T	T
T	F	F
T	T	T

5.

$\sim p$	$\sim q$	$\sim q \Rightarrow \sim p$
F	F	T
F	T	F
T	F	T
T	T	T

6.

p	$\sim p$	$p \Rightarrow \sim p$
T	F	F
F	T	T

7.

$\sim (p \Rightarrow q)$
F
T
F
F

9.

p	q	$p \Rightarrow q$	$(p \Rightarrow q) \Rightarrow q$
T	T	T	T
T	F	F	T
F	T	T	T
F	F	T	F

Exercise Set 7.5

2. (b)

p	q	$p \wedge q$	$(p \wedge q) \Rightarrow p$
T	T	T	T
T	F	F	T
F	T	F	T
F	F	F	T

(f)

p	q	$\sim p$	$p \Rightarrow q$	$\sim p \vee q$	$\Leftrightarrow$
T	T	F	T	T	T
T	F	F	F	F	T
F	T	T	T	T	T
F	F	T	T	T	T

Exercise Set 7.6

2. (a) Valid.

(b) Invalid.

(c) Valid.

(d) Valid.

(e) Valid.

Exercise Set 8.1

2. No.

3. No.

Exercise Set 8.2

4. (a) No. (b) Yes. (c) Yes. (d) Yes. (e) Yes. (f) Yes. (g) Yes.
(h) No. (i) Yes, if only lines considered are h-lines and c-lines.

5. 8.8 and 8.12: When two lines are cut by a transversal, they are parallel if and only if a pair of alternate angles are congruent.

8.9 and 8.13: When two lines are cut by a transversal, they are parallel if and only if a pair of corresponding angles are congruent.

8.10 and 8.14: When two lines are cut by a transversal, they are parallel if and only if a pair of alternate exterior angles are congruent.

8.11 and 8.15: When two lines are cut by a transversal, they are parallel if and only if a pair of interior angles whose interiors are on the same side of the transversal are supplementary.

Exercise Set 9.1

1. (a) 21 sq ft.   (c) $\frac{5}{6}$ sq ft (120 sq in.).
   (b) 5.58 sq cm.   (d) $\frac{14}{3}$ in.

2. (a) 1.53 sq mm.   (b) $2\frac{1}{2}$ sq in.   (c) $7\frac{1}{2}$ sq ft.
   (d) 4.365 cm.

3. (a) 810 sq mm.   (b) $3\frac{1}{2}$ sq ft.   (c) 10 yd.   (d) 6 cm.

4. (a) Approximately 13.5 sq units.
   (b) Approximately 43.0 sq units.

7. (a) Area quadrupled.   (c) Area tripled.   (e) Area quadrupled.
   (b) Area halved.   (d) Area doubled.   (f) Area unchanged.

Exercise Set 9.2

1.   *Perimeter*   *Area*
   (a) 12 cm $4\sqrt{3}$ sq cm
   (b) 18 in. $9\sqrt{3}$ sq in.
   (c) 6.3 mm $\frac{4.41}{4}\sqrt{3}$ sq mm
   (d) $18\sqrt{3}$ ft $27\sqrt{3}$ sq ft
   (e) $3s$ linear units $\frac{s^2}{4}\sqrt{3}$ sq units.

2. $4\sqrt{3}$

3.   *Area*   *Perimeter*
   (a) Approx. 110.0 sq yd   40 yd
   (b) Approx. 0.9675 sq m   3.75 m
   (c) $\frac{5}{4}s^2$ • tan 54 sq units $5s$

4.   *Area*   *Perimeter*
   (a) $\frac{243}{2}\sqrt{3}$ sq in.   54 in.
   (b) $\frac{8}{3}\sqrt{3}$ sq ft   8 ft
   (c) $27\sqrt{3}$ sq mm $18\sqrt{2}$ mm
   (d) $\frac{3}{2}s^2\sqrt{3}$ sq units $6s$ units

7. 180.

8. $103\frac{1}{2}$ sq units.

Exercise Set 9.3

1.   *Perimeter*   *Area*
   (a)   5.00–5.40 ft   1.5625–1.8225 sq ft
   (b) 49.820–49.860 cm   155.127025–155.376225 sq cm
   (c)   15.60–16.00 in.   14.6475–15.4375 sq in.
   (d)   0.300–0.340 m   0.003125–0.004725 sq m
   (e)   16.35–16.65 in. $7.425625\sqrt{3}$–$7.700625\sqrt{3}$ sq in.

Exercise Set 10.1

2. (a) *F*.   (e) *T*.   (i) *T*.   (m) *F*.
   (b) *F*.   (f) *T*.   (j) *F*.   (n) *T*.
   (c) *F*.   (g) *F*.   (k) *T*.
   (d) *T*.   (h) *F*.   (l) *T*.

Exercise Set 10.2

2. (a) $\overleftrightarrow{BC}$.   (c) $\overleftrightarrow{C}$.   (e) *A*.   (g) $\overleftrightarrow{AB}$.

   (b) $\varnothing$.   (d) $\overleftrightarrow{CD}$.   (f) *D*.   (h) *B*.

3. Yes. Smaller.

4. Yes. 180.

5. Yes. 180.

6. Yes. 360.

Exercise Set 10.3

2. 60; 90; 108; 120.

3. 3.

4. 5.

5. 3, 4, or 5.

6. 3.

7. 3.

8. 5.

9. (a) $3V$.　　　　(b) V.　　　　(c) $3V/2$.　　　　(d) $V=4$.
   (e) $F=4$; $E=6$.

10. (1) (a) $4V$.　　(b) $4V/3$.　　(c) $2V$.　　(d) $V=6$.
    (e) $F=8$; $E=12$.

   (2) (a) $5V$.　　(b) $5V/3$.　　(d) $5V/2$.　　(d) $V=12$.
   (e) $F=20$; $E=30$.

   (3) (a) $3V$.　　(b) $3V/4$.　　(c) $3V/2$.　　(d) $V=8$.
   (e) $F=6$; $E=12$.

   (4) (a) $3V$.　　(b) $3V/5$.　　(c) $3V/2$.　　(d) $V=20$.
   (e) $F=12$; $E=30$.

Exercise Set 10.4

1.

V	E	F
6	9	5
8	12	6
10	15	7
12	18	8
16	24	10

3. Quadrangular prism.

6. (a) T.　　(c) F.　　(e) F.　　(g) T.　　(i) T.
   (b) F.　　(d) F.　　(f) T.　　(h) F.　　(j) T.

Exercise Set 10.5

1. (a) 47.84 cu in.; 83.92 sq in.
      8.4 cu ft; 28.6 sq ft
      1320 cu cm; 746 sq cm
      $2\frac{7}{8}$ in.; 176 sq in.
      21 ft 4 in.; $186\frac{7}{12}$ sq ft
   (b) 198 cu mm; 95.81 cu yd; 16 sq in.
   (c) 52 cu cm; 24 sq ft; 2.1 in.

2. (a) $25\sqrt{3}$ sq in.; $125\sqrt{2}/12$ cu in.
   (b) $9\sqrt{3}$ sq in.; $\frac{9}{4}\sqrt{2}$ cu in.

3. (a) 2.16 sq cm; 0.216 cu cm
   (b) 12 sq in.; $2\sqrt{2}$ cu in.

4. 384.

5. 400.

10.

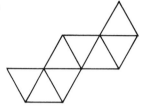
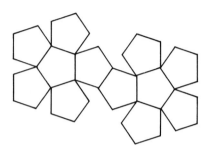

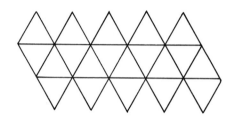

Exercise Set 11.1

2. It is perpendicular.

5. 2.

6. Infinitely many.

7. Infinitely many.

8. 2.

Exercise Set 11.2

1. No.

2. Yes.

7. No. . . . Regular polygons.

Exercise Set 11.3

1. They are supplementary.

2. It is a rectangle.

6. Equilateral—60; isosceles right—90.

Exercise Set 11.4

1. (a) 14   28   88   616
        1.4   2.8   8.8   6.16
        $1\frac{3}{4}$ $3\frac{1}{2}$ 11 $9\frac{5}{8}$
        4.2   8.4   26.4   55.44
   (b) 4   8 $\frac{1408}{7}$ $\frac{4232}{21}$
        3   6 $\frac{792}{7}$ $\frac{792}{7}$
        $\frac{7}{44}$ $\frac{7}{22}$ $\frac{7}{22}$ $\frac{49}{2904}$

2. Approximately 25,133 miles.

4. (a) One-half as large.   (b) Yes.

Exercise Set 11.5

2. $2\pi rh$.

3. If $s = m(\overline{PQ})$, then lateral area $= \pi rs$.

4. (a) 3   2 12π 30π 18π
       2   2 8π 16π 8π
       3   4 24π 42π 36π

   (b) 3   4 15π 24π 12π
       2   2 $4\pi\sqrt{2}$ $(4\sqrt{2}+4)\pi$ $8\pi/3$
       3   2 $3\pi\sqrt{13}$ $(3\sqrt{13}+9)\pi$ 6π

5. Four times the height of the cone.

6. $\frac{\pi}{3}h(R^2 - r^2)$.

Exercise Set 13.1

1. A: $(5, -3)$; B: $(2, -5)$; C: $(3, 0)$; D: $(6, 1)$; E: $(0, -5)$; F: $(-4, -3)$;
   G:$(-7, -6)$; H:$(-1, 3)$; J:$(-5, 5)$.

2. $(0, 0)$.

3. 0; 0.

4. $(0, 9)$.

5. Q: perimeter—16 units; area—7 sq units
   S: perimeter—10 units; area—6 sq units

M: perimeter—18 units; area—20 sq units.

6. D: $(5, -3)$ perimeter—26 units; area—40 sq units.

7. Q $(6, -4)$ perimeter—16 units; area—9 sq units.

Exercise Set 13.2

1. (a) $\overleftrightarrow{PQ}$: $x + y + 1 = 0$.   (f) $\overleftrightarrow{XY}$: $3x + 10y + 15 = 0$

   (b) $\overleftrightarrow{AB}$: $10x - 3y + 11 = 0$.   (g) $\overleftrightarrow{JK}$: $x = 1$.

   (c) $\overleftrightarrow{MN}$: $x + y - 4 = 0$.   (h) $\overleftrightarrow{LM}$: $3x + 4y - 16 = 0$.

   (d) $\overleftrightarrow{EF}$: $x - y - 2 = 0$.   (i) $\overleftrightarrow{BG}$: $3x + 4y - 31 = 0$.

   (e) $\overleftrightarrow{RT}$: $y = 2$.   (j) $\overleftrightarrow{SV}$: $8x - 6y - 15 = 0$.

2. Parallel: $\overleftrightarrow{PQ}$ and $\overleftrightarrow{MN}$; $\overleftrightarrow{LM}$ and $\overleftrightarrow{BG}$.

   Perpendicular: $\overleftrightarrow{PQ}$ and $\overleftrightarrow{EF}$; $\overleftrightarrow{MN}$ and $\overleftrightarrow{EF}$; $\overleftrightarrow{RT}$ and $\overleftrightarrow{JK}$; $\overleftrightarrow{AB}$ and $\overleftrightarrow{XY}$; $\overleftrightarrow{LM}$ and $\overleftrightarrow{SV}$; $\overleftrightarrow{BG}$ and $\overleftrightarrow{SV}$.

3. $m_{MN} = m_{PH} = 2$.

4. $m_{AB} = m_{CD} = 1$; $m_{AD} = m_{BC} = 3$.

5. $m_{WX} = m_{YZ} = \frac{4}{3}$; $m_{WZ} = m_{XY} = -\frac{3}{4}$.

6. $m_{RT} = -\frac{21}{2}$; $m_{ST} = \frac{2}{21}$.

7.

Slope	x-intercept	y-intercept
(a) $\frac{3}{2}$	$-\frac{7}{3}$	$\frac{7}{2}$
(b) $-\frac{1}{5}$	3	$\frac{3}{5}$
(c) $-\frac{2}{3}$	-2	$-\frac{4}{3}$
(d) $\frac{3}{2}$	$-\frac{5}{6}$	$\frac{5}{4}$
(e) -1	0	0
(f) -1	-1	-1
(g) 1	0	0
(h) $\frac{9}{16}$	32	-18
(i) 1	3	-3
(j) 5	$-\frac{11}{5}$	11

8. Parallel: a and d; e and f; g and i.

   Perpendicular: a and c; d and c; b and j; e and g; e and i; f and g; f and i.

Exercise Set 13.3

1. (a) $2\sqrt{2}$. (b) $\sqrt{109}$. (c) $3\sqrt{2}$. (d) $2\sqrt{2}$. (e). 10. (f) $\sqrt{109}$.
(g) 4. (h) $\frac{5}{2}$. (i) $\frac{5}{2}$. (j) 10.

2. (a) $(2, -3)$. (d) $(1, -1)$. (g) $(1, -5)$. (j) $(\frac{15}{4}, \frac{5}{2})$.
(b) $(-\frac{1}{2}, 2)$. (e) $(1, 2)$. (h) $(1, \frac{13}{4})$.
(c) $(1\frac{1}{2}, 2, \frac{1}{2})$. (f) $(0, -1\frac{1}{2})$. (i) $(0, \frac{31}{4})$.

3. (a) $4\sqrt{5} + 6\sqrt{2}$. (c) $10 + 7\sqrt{2}$.
(b) $\sqrt{13} + 82 + \sqrt{65}$. (d) $12 + 6\sqrt{5}$.

4. Isosceles: a, c, and d.
Equilateral: none.

5. $m_{AB} = m_{CD} = \frac{1}{8}; m_{AC} = m_{BC} = -\frac{7}{4}; m(\overline{AB}) = m(\overline{BC}) = \sqrt{65}$.

6. $[m(\overline{MN}) = 5; m(\overline{NQ}) = 5; m(\overline{MQ}) = 7\sqrt{2}] \Rightarrow m(\overline{MN}) + m(\overline{NQ}) > m(\overline{MQ})$.

7. from R: $\sqrt{52}$;
from S: 5;
from T: $\sqrt{37}$.

9. $x^2 + y^2 = r^2$

10. (a) $x^2 + (y + 5)^2 = 1$.
(b) $(x - 1)^2 + (y - 4)^2 = 25$.
(c) $(x + 2)^2 + y^2 = 4$.
(d) $(x + 1)^2 + (y + 3)^2 = 4$.
(e) $(x - 2)^2 + (y + 1)^2 = 9$.

INDEX